Lecture Notes in Mathematics

2022

Editors:
J.-M. Morel, Cachan
B. Teissier, Paris

Editors *Mathematical Biosciences Subse*
P.K. Maini, Oxford

For further volumes:
http://www.springer.com/series/304

Ingemar Nåsell

Extinction and Quasi-stationarity in the Stochastic Logistic SIS Model

 Springer

Ingemar Nåsell
Royal Institute of Technology
Mathematics
10044 Stockholm
Sweden
ingemar@kth.se

ISBN 978-3-642-20529-3   e-ISBN 978-3-642-20530-9
DOI 10.1007/978-3-642-20530-9
Springer Heidelberg Dordrecht London New York

Lecture Notes in Mathematics ISSN print edition: 0075-8434
                             ISSN electronic edition: 1617-9692

Library of Congress Control Number: 2011930820

Mathematics Subject Classification (2000): 60-J28; 92-D30

© Springer-Verlag Berlin Heidelberg 2011
This work is subject to copyright. All rights are reserved, whether the whole or part of the material is
concerned, specifically the rights of translation, reprinting, reuse of illustrations, recitation, broadcasting,
reproduction on microfilm or in any other way, and storage in data banks. Duplication of this publication
or parts thereof is permitted only under the provisions of the German Copyright Law of September 9,
1965, in its current version, and permission for use must always be obtained from Springer. Violations
are liable to prosecution under the German Copyright Law.
The use of general descriptive names, registered names, trademarks, etc. in this publication does not
imply, even in the absence of a specific statement, that such names are exempt from the relevant protective
laws and regulations and therefore free for general use.

Cover design: deblik, Berlin

Printed on acid-free paper

Springer is part of Springer Science+Business Media (www.springer.com)

Preface

Mathematical modeling in population biology aims to give insight into the dynamics of both single populations and multiple interacting populations. Fundamental questions are of qualitative nature. They concern the long-term survival or extinction of any given population or subpopulation. Work in this area has for a long time followed two different mathematical branches: deterministic and stochastic. The deterministic models have given powerful results by responding to fundamental qualitative questions concerning survival and extinction. At the same time, it is well known that the deterministic models suffer from a serious weakness: the state-space used is continuous, while counts of actual populations are always given by nonnegative integers. This weakness is avoided by using stochastic models with discrete state space. The question therefore arises if the qualitative results that hold in the deterministic setting have counterparts in the more realistic stochastic version of the model. Another way of formulating this question is to ask whether the threshold results that hold in the deterministic setting have counterparts in the stochastic world.

There is no easy or general answer to this question. We approach it by presenting an analysis of the so-called SIS model that is used to study the spread of infection without immunity in a constant population. This is one of the simplest models where this investigation is meaningful. Our methods may be useful in more complicated and more realistic models. The deterministic version of the SIS model has a threshold where a parameter denoted by R_0 takes the value 1. This means that an endemic infection level will establish itself if $R_0 > 1$ and the initial proportion of infected individuals is positive, while any infection will ultimately disappear if $R_0 < 1$. The stochastic version of the SIS model takes the form of a birth–death process with an absorbing state at the origin. Here, any infection will ultimately disappear for all values of R_0. We investigate the long-term behavior of this model by studying both the time to extinction and the so-called quasi-stationary distribution, which is a stationary distribution, conditional on not being absorbed at the origin. Its analysis is more intricate than the analysis of an ordinary stationary distribution of a birth–death process. In particular, explicit solutions are not available. Therefore, the emphasis is on deriving approximations.

Several approximation steps are required to derive satisfactory approximations of the quasi-stationary distribution $\mathbf{q}$ of the SIS model. These approximation steps have been developed gradually over time from 1978 till 2011. The first step was taken by Cavender (1978) [16]. He introduced an auxiliary process as a related birth–death process without an absorbing state. This auxiliary process has a stationary distribution $\mathbf{p}^{(0)}$ that can be determined explicitly. Cavender showed that it serves as a lower bound of $\mathbf{q}$ in the sense of stochastic ordering: $\mathbf{p}^{(0)} \prec_{ST} \mathbf{q}$.

The next step was taken by Kryscio and Lefèvre (1989) [39]. They introduced an additional auxiliary process as a second related birth–death process without an absorbing state. Again, its stationary distribution $\mathbf{p}^{(1)}$ can be determined explicitly. Also, Kryscio and Lefèvre made the important conjecture that this stationary distribution provides an upper bound of the quasi-stationary distribution $\mathbf{q}$ in the sense of stochastic ordering: $\mathbf{q} \prec_{ST} \mathbf{p}^{(1)}$.

After this, Nåsell (1996) [44] proceeded to derive approximations of the stationary distributions $\mathbf{p}^{(0)}$ and $\mathbf{p}^{(1)}$. He searched for asymptotic approximations as the total population size N became large. He showed that qualitatively different results could be established in three different parameter regions. Two of these were counterparts to the two regions $R_0 < 1$ and $R_0 > 1$ that hold in the deterministic case, while a third region appeared as a transition region near $R_0 = 1$. Nåsell also identified a map Ψ that could be used to determine $\mathbf{q}$ numerically by iteration. He used this map to derive explicit approximations of $\mathbf{q}$ by applying one iteration step to the approximation $\mathbf{p}^{(1)}$. The resulting approximation is ad hoc, since it was not then clear in what sense an approximation of $\Psi\left(\mathbf{p}^{(1)}\right)$ is an approximation of the quasi-stationary distribution $\mathbf{q}$ itself.

The map Ψ was also used by Ferrari, Kesten, Martínez, and Picco (1995) [27], although this publication was not available when the 1996 paper by Nåsell was written. Ferrari et al. showed that the map Ψ was important for the quasi-stationary distribution $\mathbf{q}$ in the sense that $\lim_{i \to \infty} \Psi^i(\mathbf{p}) = \mathbf{q}$ holds for arbitrary distributions $\mathbf{p}$, and also that $\Psi(\mathbf{q}) = \mathbf{q}$.

The conjecture by Kryscio and Lefèvre that $\mathbf{p}^{(1)}$ is an upper bound of $\mathbf{q}$ in the sense of stochastic ordering remained an open problem until it was settled by Clancy and Pollett (2003) [18]. In their proof of this conjecture, they made use of the map Ψ. They also established an important theorem that showed that the map Ψ preserves a certain ordering between probability vectors.

In this monograph, we derive approximations of some of the distributions $\Psi\left(\mathbf{p}^{(1)}\right)$, $\Psi\left(\mathbf{p}^{(0)}\right)$, and $\Psi^i\left(\mathbf{p}^{(1)}\right)$ in the three parameter regions. Using the theorem established by Clancy and Pollett, we are able to show that these approximations actually give approximations of the quasi-stationary distribution $\mathbf{q}$ itself.

The final step in the derivation of approximations of the quasi-stationary distribution is also taken in this monograph, where we derive approximations that are uniformly valid across all three of the parameter regions.

We also give approximations of the time to extinction for two specific initial distributions. The first case is used to study extinction time for an established infection, and the second one is used to study establishment of an infection. For the first case, we let the initial distribution be equal to the quasi-stationary distribution. The time to extinction then has an exponential distribution whose expectation is determined by the quasi-stationary distribution. Its approximation is therefore found from the approximation of the quasi-stationary distribution described above. For the second case, we consider the case when initially one infective individual is present. The expected time to extinction can then be determined from the distribution $\mathbf{p}^{(0)}$, for which an approximation has been derived, as described above.

The monograph is written for a reader who has a good working knowledge about birth–death processes. An introduction is given to the less well-known concept of quasi-stationarity, and to its relation with extinction times. Heavy use is made in the monograph of ideas from the area of asymptotic approximation. An effort is made to present these ideas in a simple way, since it is an area of applied mathematics that is less well known to the stochastic community.

I am grateful to Joshua Ross and to three anonymous referees for careful reading of the manuscript and for making valuable comments that have improved the presentation of the results considerably.

Stockholm   *Ingemar Nåsell*
March 2011

Contents

Chapter 1
Introduction

This monograph is devoted to an analysis of a classical mathematical model in population biology, known as the stochastic logistic SIS model. It serves as a model both for the spread of an infection that gives no immunity and for density dependent population growth, and it also appears as an important special case of a contact process that accounts for spatial influences. These three interpretations of the model are further discussed in Chap. 2. In that chapter it is also shown that the SIS model is just one out of a whole family of logistic models. The model is used outside population biology in areas such as the spread of rumours, the spread of technical innovations, and the theory of chemical reactions. The deterministic version of this model takes the form of a nonlinear differential equation that can be solved explicitly, and where a bifurcation phenomenon appears that corresponds to a very powerful qualitative so-called threshold result. We are mainly concerned with the stochastic version of this model, and with establishing counterparts in the stochastic model to the threshold result that holds for the deterministic model.

The deterministic version of the logistic model goes back to Verhulst [74], while the first studies of its stochastic version are due to Feller [26], Bartlett [11], and Weiss and Dishon [76]. The interest in this model grew slowly at first. One sign of this is that the model was not discussed in the influential book on models in mathematical epidemiology by Bailey [7], although references to the papers by Feller and by Weiss and Dishon were included. The model has since then appeared in several contexts. Bartholomew [9] has applied it to study the transmission of rumours, Oppenheim et al. [56] use it as a model for chemical reactions, Cavender [16] uses it as an example of a birth-and-death process, Norden [53] describes it as a stochastic logistic model, while Kryscio and Lefèvre [39], Nåsell [44, 45, 47, 49], and Andersson and Djehiche [4] return to the epidemic context. Kryscio and Lefèvre summarize and extend the work of the previous authors. Cavender [16] and Kryscio and Lefèvre [39] introduce two very useful auxiliary processes discussed below. Nåsell [44] provides extensions of these results. He introduces the important transition region (see below) into the study. Further improvements given in this monograph are based on the papers by Ferrari et al. [27] and by Clancy and Pollett

I. Nåsell, *Extinction and Quasi-stationarity in the Stochastic Logistic SIS Model*,
Lecture Notes in Mathematics 2022, DOI 10.1007/978-3-642-20530-9_1,
© Springer-Verlag Berlin Heidelberg 2011

[18]. Additional work on this and similar models are reported by Grasman [31], Ovaskainen [57], Newman et al. [51], Doering et al. [24], Assaf et al. [6], and Cairns et al. [15].

The stochastic model that we deal with takes the form of a birth–death process with discrete state space and continuous time. Using the language of epidemic modelling, we find that the state of the process gives the number of infected individuals as a function of time in a constant population of N individuals. It takes integer values from zero up to N. The model accounts for two changes in state, corresponding to infection and recovery. The hypotheses of the model specify the rates at which these two changes occur. The model properties can be derived from a linear system of differential equations for its state probabilities, called Kolmogorov equations. The number of variables in this system of equations equals the total number of states, which is $N + 1$.

It is very useful for the analysis to consider the deterministic version of this stochastic model. It can be derived in two different ways. The first way of deriving it, which is common among deterministic modellers, is to interpret the hypothesized transition rates deterministically. This leads immediately to a non-linear differential equation for the number or the proportion of infected individuals. Its solution gives this proportion as a function of time. We see here one important difference between stochastic and deterministic models: The state variable of the latter is not limited to integer values, since it appears in the model as a function of time that is differentiable and therefore continuous.

The second way of deriving the deterministic version of the model is as an approximation of the stochastic one. We are lead to the same differential equation as above after scaling with the population size N and letting N approach infinity. One consequence of this is that the parameter N disappears from the scene in the deterministic setting. The influence of N is important in the stochastic setup, and absent in the deterministic one. The fact that the deterministic model is an approximation of the stochastic one raises the question if the approximation is acceptable. Apparently, acceptability requires the population size N to be sufficiently large. But it is important to agree on what criteria should be used in judging such acceptability. The only way to find out is to do a full analysis of the stochastic model. This is the purpose of the present monograph for the stochastic SIS model. The main mathematical tools that we shall use in this work are taken from the area of asymptotic analysis. We shall develop approximations of the various quantities that are of interest in the stochastic model as N approaches infinity. We strive for approximations that are asymptotic, but we shall not always succeed.

Deterministic modelling has been very successful in many areas of population biology. The important results are qualitative in nature, and are derived from nonlinear deterministic models with bifurcation. These results respond to qualitative questions about the survival or extinction of specific populations. It is not clear what the counterpart to these qualitative results is in the stochastic version of the model, especially since the reason for the qualitative results is nonlinearity in the deterministic formulation, while the nonlinearity is absent in the stochastic formulation. We shall illuminate this question in the simple case of the stochastic logistic SIS model.

It is noteworthy that deterministic and stochastic models disagree qualitatively with regard to the extinction phenomenon. The deterministic version of the SIS model has a threshold. This means that the population of infected individuals is predicted to go extinct if it lies below the threshold, and that it will persist indefinitely if it lies above the threshold. (We note that the threshold is a point in a parameter space. We say that a population lies above (below) the threshold if a certain parameter lies above (below) its threshold value.) In stark contrast to this, we find that the stochastic model predicts that the population of infected individuals will ultimately go extinct as time goes on, regardless of where the population is situated with regard to the deterministic model threshold. However, the time to extinction turns out to be very different above and below the deterministic model threshold; it can be really long above the threshold, in some cases exceeding the age of the universe, while it is short below it. The time to extinction can therefore be used as a kind of counterpart in the stochastic version of the model to the qualitative result represented by the threshold that can be established for the deterministic version of the model. Our study of the time to extinction for the stochastic SIS model will confirm this.

The fact that deterministic and stochastic models disagree qualitatively as just mentioned may appear inconsistent with the property that the deterministic model is an approximation of the stochastic one, but that is not so. As shown in Andersson and Britton [3], the deterministic model solution is an approximation of the solution for the stochastic model only on finite time intervals. Thus, the approximation does not necessarily hold in the limit when time approaches infinity. A consequence of this is that the endemic infection level predicted by a deterministic model corresponds to the quasi-stationary distribution rather than to the stationary distribution in a stochastic model with an absorbing state, as is the case in the model that is studied here.

The threshold phenomenon can be described by a partition of parameter space into regions where model properties differ qualitatively. For the deterministic model it leads to two regions, one above threshold, and the other one below threshold. It is customary to introduce a parameter that is denoted by R_0 and referred to as the basic reproduction ratio, and with the property that the threshold is identified by $R_0 = 1$. The situation is different in the stochastic model. The fact that the extra parameter N is present in the stochastic model has the consequence that this model has three parameter regions where qualitatively different results occur. Two of them correspond roughly to the two regions $R_0 > 1$ and $R_0 < 1$ that are present in the deterministic setup, while the third one is a transition region between the two that appears near the threshold where $R_0 = 1$. The stochastic model parameter region that corresponds to the region where $R_0 > 1$ in the deterministic model can be described in two ways. The first description is purely formal, and based on the fact that we use concepts from the area of asymptotic analysis. The parameter region is then defined by the requirement that $R_0 > 1$ as $N \to \infty$. However, from a practical standpoint this means that $R_0 - 1$ must be strictly positive for any finite value of N. We shall describe this by saying that R_0 is distinctly above one. The second description of the parameter region where R_0 is distinctly above one is in terms of approximate boundary values for R_0 that depend on N. These boundary values decrease toward the value one as N increases toward infinity. Similar descriptions apply for the parameter region where R_0 is distinctly below one. The width of the transition region goes toward zero as N becomes large. In the limit as N reaches infinity, the transition region becomes empty, as it is in the deterministic version of the model.

The various manifestations of the stochastic logistic model that we are concerned with all behave in qualitatively different ways in these three parameter regions. Therefore, by necessity we are led to consider separate derivations of our results in each of these regions.

The qualitative difference between deterministic and stochastic models mentioned above has a consequence that is important for our analysis. The counterpart to a stable stationary solution of the deterministic model is not a stationary distribution of the stochastic model, but instead a so-called quasi-stationary distribution. The stationary distribution of the stochastic model is degenerate. It is reached when extinction occurs. However, before extinction it will in many cases be true that the distribution is practically constant. This distribution, called the quasi-stationary distribution, can mathematically be defined as a stationary distribution, conditional on non-extinction. It has a very desirable property from a modelling standpoint. If the system that we are studying has been running for a long time, and if the only thing that we know about it is that it has not reached extinction, then we can conclude that the quasi-stationary distribution is the likely distribution of the state variable. This is one reason for our interest in the quasi-stationary distribution. Another reason is that knowledge about this distribution also gives us information about the remaining time to extinction from this distribution. This is a mathematical result that we make use of in the analysis of the extinction time for the model that we study. Because of these two properties, the quasi-stationary distribution plays a central role in our study of the stochastic SIS model. A bit of history of the concept of quasi-stationarity is given by Pollett [61]. It shows that this concept originated with the two papers by Yaglom [78] and by Bartlett [10]. Even though the Yaglom paper preceded that of Bartlett, it appears that Bartlett's ideas about quasi-stationarity were original and unaffected by Yaglom. Basic theoretical results concerning quasi-stationary distributions for continuous-time Markov Chains with finite state space were later derived by Darroch and Seneta [21]. It is likely that they were inspired by Bartlett, as indicated in Nåsell [50].

It turns out to be impossible to find explicit expressions for the quasi-stationary distribution for any population model with density dependence. Progress therefore rests on finding good approximations. All approximations of the quasi-stationary distribution of the SIS model that we consider in this monograph are based on two auxiliary processes introduced by Cavender [16] and by Kryscio and Lefèvre [39]. These processes are birth–death processes whose transition rates are similar to those of the SIS model itself. An important difference is that the origin has been removed from their state spaces. Thus, they lack absorbing states, and have the same state space as the quasi-stationary distribution. This means that they have non-degenerate stationary distributions that can be determined explicitly. These stationary distributions, called $\mathbf{p}^{(1)}$ and $\mathbf{p}^{(0)}$, both serve as approximations of the quasi-stationary distribution.

In an early study of the SIS model, Nåsell [44] noted that the components of the quasi-stationary distribution satisfy a certain implicit relation. Moreover, this relation suggested that it should be possible to solve for the quasi-stationary distribution with the aid of iteration, where the stationary distributions $\mathbf{p}^{(1)}$ and $\mathbf{p}^{(0)}$ suggested themselves as natural starting points. Indeed, it turned out that a numerical method based on this iteration scheme seemed to converge. It was therefore natural to imagine that an approximation of the quasi-stationary distribution would be produced by taking just one iteration step. However, it is not clear in what sense an asymptotic approximation after one iteration step approximates the quasi-stationary distribution itself. This weakness with the indicated method can now be resolved.

The iteration method can be described by giving a sequence of iterates, where each one is produced by applying a certain map Ψ to the previous one. This map will be defined in Chap. 3. It was independently defined and studied in the theoretical work by Ferrari et al. [27]. They showed that the sequence of iterates converged to the quasi-stationary distribution for arbitrary initial distributions. A further important property of this map was shown by Clancy and Pollett [18]. They proved that it preserves what is called likelihood ratio ordering. A proof is given in Sect. 3.6. By using this result, we are able to derive asymptotic approximations of the quasi-stationary distribution itself, and not only of the iterates, in the two parameter regions where R_0 is distinctly larger than one and distinctly smaller than one. We shall also provide an approximation of the quasi-stationary distribution in the transition region, but we do not claim that it is asymptotic.

The concept of quasi-stationarity is important for many models in population biology. It is therefore desirable to have access to methods that give information about the quasi-stationary distribution for such models. The methods that are developed in the present study are of value for other models. This is particularly true for univariate logistic models, for which preparations are made in Chap. 3. But it holds also for the two bivariate models where already some work on the quasi-stationary distribution has been reported, namely the Ross malaria model and the classical SIR model with demography, used for studying childhood infections, see Nåsell [42, 46, 50].

There are two aims of the present monograph. One of them is, as mentioned above, to give a useful approximation of the quasi-stationary distribution and of the time to extinction for the stochastic SIS model. The second aim is to describe the methods that we have developed for deriving such approximations in such a way that the reader gains an insight that is useful in work on related stochastic models.

The rest of the monograph is disposed as follows:

Chapter 2 is devoted to model formulation. We give three different population biology situations that lead to similar mathematical models. We show in particular that the SIS model is one out of a whole class of logistic models.

Chapter 3 gives important stochastic process background for our study, to a large part in a more general setting than that provided by the SIS model. We deal with the quasi-stationary distribution, the time to extinction, and the two auxiliary processes introduced by Cavender [16] and by Kryscio and Lefèvre [39]. We show that the stationary distributions $\mathbf{p}^{(0)}$ and $\mathbf{p}^{(1)}$ of the auxiliary processes are important both for the quasi-stationary distribution and for the time to extinction. Also, we define the map Ψ mentioned above. It is a map between discrete distributions with the property that the quasi-stationary distribution is the unique fixed point of this map. We discuss also concepts and results of stochastic ordering. An important conjecture concerning stochastic ordering for the SIS model pronounced by Kryscio and Lefèvre in 1989 is described. A proof of this conjecture for the SIS model was given by Clancy and Pollett [18]. Because of its importance, it is included here.

We return to the stochastic SIS model in Chap. 4. We give explicit expressions for the stationary distributions $\mathbf{p}^{(1)}$ and $\mathbf{p}^{(0)}$ of the two auxiliary processes introduced by Cavender [16] and by Kryscio and Lefèvre [39]. As mentioned above, these stationary distributions were originally introduced as approximations of the quasi-stationary distribution of the SIS model. In our approach, they are starting points for approximations that lead to the main result that we present, namely a uniform approximation of the quasi-stationary distribution, valid over all three parameter regions. Chapter 4 also contains numerical illustrations that show that the two stationary distributions $\mathbf{p}^{(1)}$ and $\mathbf{p}^{(0)}$ do not provide sufficiently good approximations of the quasi-stationary distribution $\mathbf{q}$.

Many of the various approximations for the SIS model that are developed in later chapters are based on approximations involving the normal distribution. This type of result is of course completely independent of the particular model that we study. We derive the approximations concerning the normal distribution that we need in Chap. 5. They include asymptotic approximations of sums of normal densities, and of sums of reciprocals of normal densities. Even though the normal distribution has been studied extensively, it appears that these approximation results are new. The results in this chapter are likely to be highly useful in the search for approximations of quasi-stationary distributions of other stochastic models.

Approximations of the stationary distribution $\mathbf{p}^{(1)}$ are derived in Chaps. 6 and 7. The first of these two chapters is devoted to derivations, while the second chapter summarizes the results. Separate approximations are given in each of the three parameter regions. Furthermore, each approximation of the probability $p_n^{(1)}$ is valid only in a restricted interval of n-values. This interval is indicated in each case.

Very similarly, we give approximations of the stationary distribution $\mathbf{p}^{(0)}$ in Chaps. 8 and 9. Again, separate results are given in each of the three parameter regions, and the results hold in restricted n-intervals.

Approximations of images under the map Ψ of the stationary distributions $\mathbf{p}^{(1)}$ and $\mathbf{p}^{(0)}$ are derived in Chap. 10. We give also approximations of the images of $\mathbf{p}^{(1)}$ after multiple applications of the map Ψ. These results are used in Chap. 11 to derive approximations of the quasi-stationary distribution in each of the three parameter regions, again in properly restricted n-intervals.

The expected time to extinction from quasi-stationarity and from the state one can be determined mathematically from the quasi-stationary distribution $\mathbf{q}$ and from the stationary distribution $\mathbf{p}^{(0)}$, respectively. Approximations are given in Chap. 12.

As can be expected, the approximations take different forms in the three parameter regions.

The need to present results separately in three different parameter regions is somewhat unsatisfactory and unelegant. We respond to this by deriving uniform results that are valid over all three parameter regions. Uniform approximations of the two stationary distributions $\mathbf{p}^{(1)}$ and $\mathbf{p}^{(0)}$ and of the quasi-stationary distribution $\mathbf{q}$ are given in Chap. 13. This chapter also contains uniform approximations of the expected times to extinction from the quasi-stationary distribution and from the state one. All these results are new. The uniform approximations are improvements over the results that have been derived separately in each of the three parameter regions.

Chapter 14 discusses thresholds for the stochastic SIS model. Early work on this model used numerical evaluations as a basis for conjecturing the threshold behavior, as described by Nåsell [43]. We show that the approximations that we have derived are consistent with slight variations of these early conjectures.

The monograph finishes with some concluding remarks in Chap. 15.

The monograph is written for a reader who has a good working knowledge about birth–death processes, including methods for their formulation and analysis. There are many text-books that deal with this area of stochastic processes. Two good introductions are the books by Allen [2] and by Taylor and Karlin [71]. Another introduction that can be recommended is the unpublished report by Schmitz [66]. It is particularly appropriate for a reader of the present monograph, since it deals with particularities of the SIS model, including a treatment of its quasi-stationary distribution. It is written in German.

We envision two different readers of this monograph. One of them is mainly interested in what can be said about the quasi-stationary distribution and the time to extinction for the SIS model, while the other one is interested not only in these results, but also wants to understand the methods that we use and to apply them in analysis of quasi-stationary distributions of other stochastic models. The reader in the first category is recommended to go directly to Chaps. 11–15, after reading the introductory chap. 2, and briefly reviewing the contents of Chaps. 3–5. In particular, he is encouraged to skip the rather technical developments that are given in the preparatory Chaps. 6, 8, and 10. The very essence of the results are contained in Sects. 13.4 and 13.5, where we give uniform approximations of both the quasi-stationary distribution and of the expected time to extinction for the SIS model. However, it is important to realize that the results of the intermediate Chaps. 6–10 are all needed in the important development in Chap. 11, where approximations of the quasi-stationary approximation are derived. These intermediate chapters are on the other hand recommended for the reader in the second category. In particular, they will acquaint him with the development of asymptotic approximations, which may be less common knowledge for many persons that work with stochastic models.

There are two appendices to this monograph. The first one gives a summary of the notation that is used, while the second one contains a number of Maple procedures in the form of a Maple module. These procedures can be used to do numerical evaluations of the various quantities that are studied. In particular, they have been used to produce the plots that are included in the monograph.

Chapter 2
Model Formulation

The mathematical model dealt with in this monograph is referred to as a stochastic logistic SIS model. In this chapter we show that there are many logistic models, and that the SIS model is just one of them, and probably the simplest one. We identify three different population biology situations that lead to similar mathematical models, namely

1. The density dependent population growth originally studied by Verhulst [74],
2. The so-called SIS model, which deals with the spread of infection without immunity in a constant population, and
3. A special case of a contact process that accounts for spatial influences.

These three situations are described in the first three sections of the chapter.

The study of stochastic logistic models was in the beginning hampered by the fact that the models in question were difficult to work with mathematically, and in particular that they could not be solved explicitly. Early workers looked for mathematical problems that could be solved exactly. Therefore, they reformulated the models to make them solvable. Unfortunately, some of the major qualitative features of the models were lost in this process. We comment on some early papers of this type in Sect. 2.4.

Section 2.5 of this chapter gives some comments on the parameter spaces for the stochastic and deterministic versions of the SIS model.

2.1 The Verhulst Model for Population Growth

The history of logistic models goes back to the deterministic model for population growth established by Verhulst [74]. His model was to a large extent forgotten and later rediscovered by Pearl and Reed [58]. The Verhulst model takes the form of a nonlinear differential equation as follows:

$$X' = f(X)X, \tag{2.1}$$

I. Nåsell, *Extinction and Quasi-stationarity in the Stochastic Logistic SIS Model*, Lecture Notes in Mathematics 2022, DOI 10.1007/978-3-642-20530-9_2, © Springer-Verlag Berlin Heidelberg 2011

where X is used to denote the population size, $f(X)$ is referred to as the per-capita growth rate, and the prime refers to the derivative with respect to time. Any situation where the per-capita growth rate is a decreasing function of the population size X is said to account for "density dependence". This reflects the fact that the available resources for population growth decrease with the population size X. Verhulst argued that it is possible to consider several choices of $f(X)$ with different mathematical form, but he settled for the mathematically simplest form, namely the one where $f(X)$ is linearly decreasing in X. This means that he studied the differential equation

$$X' = (\alpha - \beta X)X, \qquad (2.2)$$

where the parameters α and β are given, and where β is positive since $f(X)$ is a decreasing function of X. Furthermore, Verhulst studied the growth of populations in a number of countries, which means that he worked with a positive α. Thus, the original Verhulst model leads to a differential equation of the form (2.2), where both α and β are positive. However, it turns out that this equation is ecologically meaningful also when $\alpha < 0$. Therefore, we shall use the term "Verhulst equation" to refer to (2.2) with one restriction only, namely that β is positive.

The Verhulst equation is closely related to the term "logistic equation", and it appears that the use of the term "logistic" in this connection originated with Verhulst. The Verhulst equation is a nonlinear differential equation that can be solved explicitly. Verhulst gave the solution as a function of time t, and referred to it as "a logistic function" in Verhulst [75]. It was therefore natural to refer to the original Verhulst equation (2.2) as a logistic equation, and to his model as a logistic model. However, the term "logistic equation" is now used for a much larger class of equations, namely any equation that can be written as (2.1) where the per-capita growth rate $f(X)$ is a decreasing function of the population size X, or, in other words, any equation with density dependence. A large number of such deterministic logistic models is discussed by Tsoularis and Wallace [72]. This means that the Verhulst model is just one of many logistic models.

In the ecological literature it is common to reparametrize the Verhulst model by putting $r = \alpha$ and $K = \alpha/\beta$. It follows then that r and K have the same signs, since $\beta > 0$, as further discussed by Gabriel et al. [29]. The two parameters r and K are referred to as "the intrinsic growth rate" and "the carrying capacity", respectively. With this reparametrization, the Verhulst equation takes the form

$$X' = r\left(1 - \frac{X}{K}\right)X. \qquad (2.3)$$

The solution of the Verhulst differential equation in this form approaches the carrying capacity K as time approaches infinity if $X(0) > 0$ for the original Verhulst model with $r = \alpha > 0$. However, a qualitatively different result appears if $r = \alpha < 0$. In that case, the solution of the Verhulst equation approaches zero as time approaches infinity. This means that the population ultimately goes extinct. These two qualitatively different behaviors constitute the threshold behavior of the

Verhulst model. We note that the original Verhulst model, where $\alpha > 0$, does not possess a threshold.

There are many stochastic versions $\{X(t)\}$ of the deterministic Verhulst model described above. Each of them takes the form of a birth–death process with finite state space. We introduce N to denote the maximum population size, and consider birth–death processes on the state space $\{0, 1, \ldots, N\}$. The reason that there are many stochastic versions is that a given per-capita growth rate can be written as the difference between a positive per-capita birth rate and a positive per-capita death rate in many ways. These different writings lead to different stochastic processes that all have the same deterministic version. Birth and death rates for the process $\{X(t)\}$ are denoted λ_n and μ_n, respectively. They are equal to the rates of change of the transition probabilities of the process from the state n to the states $n + 1$ and $n - 1$, respectively. They are determined by the per-capita birth and death rates, multiplied by n. One set-up that is consistent with the Verhulst deterministic model is as follows:

$$\lambda_n = \lambda \left(1 - \alpha_1 \frac{n}{N}\right) n, \quad n = 0, 1, \ldots, N - 1, \quad \lambda_N = 0, \tag{2.4}$$

$$\mu_n = \mu \left(1 + \alpha_2 \frac{n}{N}\right) n, \quad n = 0, 1, \ldots, N. \tag{2.5}$$

Here, $\lambda > 0$, $\mu > 0$, $0 \le \alpha_1 \le 1$, $0 \le \alpha_2$, and $\max(\alpha_1, \alpha_2) > 0$. These restrictions assure that λ_n and μ_n are nonnegative for all n from 0 to N, as they must be. Indeed, these transition rates are strictly positive with the exceptions $\lambda_0 = 0$, $\mu_0 = 0$, and $\lambda_N = 0$. The first two of these three exceptions means that the origin is an absorbing state, while the second and third exceptions mean that the birth–death process cannot escape from the state space $\{0, 1, \ldots, N\}$. Furthermore, the density dependence of the per-capita growth rate in the original Verhulst model corresponds to density dependences in the birth rate λ_n if $\alpha_1 > 0$ and in the death rate μ_n if $\alpha_2 > 0$. The requirement that at least one of α_1 and α_2 be positive assures density dependence. The parametrization of the Verhulst model given in (2.4) and (2.5) is also given by Nåsell [47]. An early analysis of the stochastic Verhulst model is also given in this paper. It is a forerunner of the results that are given here. The model that will be studied in this monograph is the so-called SIS model described in the next section. It is the special case of the Verhulst model that corresponds to $\alpha_1 = 1$ and $\alpha_2 = 0$.

The deterministic version of the stochastic model determined by the transition rates in (2.4) and (2.5) leads to the following differential equation for the population size $X(t)$:

$$X' = \lambda_X - \mu_X = \left(\lambda - \mu - (\alpha_1 \lambda + \alpha_2 \mu) \frac{X}{N}\right) X. \tag{2.6}$$

We reparametrize by introducing the strictly positive parameter

$$R_0 = \frac{\lambda}{\mu}, \tag{2.7}$$

conventionally referred to as the basic reproduction ratio. The differential equation for X takes the special form

$$X' = -\mu(\alpha_1 + \alpha_2)\frac{X^2}{N} \tag{2.8}$$

if $R_0 = 1$, while it can be written in the form of (2.3) when $R_0 \neq 1$, with $r = \mu(R_0 - 1)$ and $K = (R_0 - 1)N/(\alpha_1 R_0 + \alpha_2)$. Verhulst's original restriction of the (2.3) to positive values of r and K is thus removed, since both r and K have the same signs as $R_0 - 1$, and therefore can take both positive and negative values. It is easy to verify that the solution of this extension of the original Verhulst logistic model changes qualitatively as the parameter R_0 passes the value one. We say that the population lies above threshold if $R_0 > 1$, and below threshold if $R_0 < 1$. As mentioned above, we find that $X(t)$ approaches $K > 0$ as $t \to \infty$ above threshold, while $X(t)$ approaches 0 as $t \to \infty$ below threshold, provided in the first case that $X(0) > 0$ and in the second case $X(0) \geq 0$. As mentioned above, we study the restriction of the Verhulst model given by $\alpha_1 = 1$ and $\alpha_2 = 0$ in this monograph. Its deterministic version obeys the same threshold behavior.

There are many models that can be described as stochastic logistic models. Each of them is a birth–death process with an absorbing state at the origin, and where the birth rate λ_n and the death rate μ_n can be written in the following forms:

$$\lambda_n = f_1(n)n, \tag{2.9}$$

$$\mu_n = f_2(n)n. \tag{2.10}$$

Here, the functions f_1 and f_2 are required to be nonnegative. Furthermore, $f_1(n)$ is constant or decreasing, while $f_2(n)$ is constant or increasing. In addition, at least one of these two functions must be nonconstant. We notice that the the ratio λ_n/μ_n is a decreasing function of n. These requirements are all consistent with the idea that a logistic model is a model that accounts for density dependence.

2.2 The SIS Model for Spread of Infection Without Immunity

We consider the spread of infection in a constant population of size N. The individuals are classified as either susceptible or infectious. The number of susceptible individuals at time t is denoted $S(t)$, and the number of infectious individuals $I(t)$. Since the population is constant, we have $S(t) + I(t) = N$. We assume that the infection does not give any immunity. This means that recovered individuals immediately become susceptible. The corresponding model is called an SIS model. This indicates that the possible transitions for any individual are from susceptible to infectious, and from infectious back to susceptible.

The population is assumed uniform and homogeneously mixing. There are contacts between all individuals in the population with a contact rate λ for each individual. The proportion I/N of all contacts for any given susceptible individual is with infective individuals. Thus, the infection rate per susceptible individual equals $\lambda I/N$. This implies that the infection rate for the whole population is $\lambda SI/N$. The recovery rate for each infective individual is constant and equal to μ. This implies that the recovery rate for the whole population equals μI.

It follows that the deterministic version of the model leads to the following differential equation for the number of infected individuals I:

$$I' = \lambda \frac{(N-I)I}{N} - \mu I. \tag{2.11}$$

As in the previous section we define the basic reproduction ratio R_0 by writing

$$R_0 = \frac{\lambda}{\mu}. \tag{2.12}$$

We are then led to the differential equation

$$I' = \mu \left(R_0 - 1 - R_0 \frac{I}{N} \right) I. \tag{2.13}$$

This equation is formally identical to the (2.6) for the population size X treated in the previous section in case $\alpha_1 = 1$ and $\alpha_2 = 0$. Thus, the deterministic version of the SIS model has a threshold at $R_0 = 1$. This means that infection that is present initially will establish itself at a positive endemic level if $R_0 > 1$, and that it will go extinct if $R_0 < 1$.

The stochastic version of the SIS model leads to a birth–death process $\{I(t)\}$ with the state space $\{0, 1, \ldots, N\}$, and with birth rate λ_n and death rate μ_n given by

$$\lambda_n = \lambda \left(1 - \frac{n}{N} \right) n, \quad n = 0, 1, \ldots, N, \tag{2.14}$$

$$\mu_n = \mu n, \quad n = 0, 1, \ldots, N. \tag{2.15}$$

The birth–death process defined in this way is the object of analysis in this monograph. As already mentioned, it can be interpreted as a special case of the Verhulst logistic model described in the previous section. The stochastic version of the SIS model was first studied by Feller [26], Bartlett [11], and Weiss and Dishon [76]. An important challenge for the analysis of the stochastic version of the model is to study the counterpart to the threshold phenomenon.

The SIS model can be regarded as the simplest logistic model. The present monograph is devoted to a detailed study of the properties of the stochastic version of this model. The methods that we use here are expected to be of value in the analysis of quasi-stationarity in the many stochastic models that can be formulated as Markov chains with absorbing states, and that are awaiting a detailed investigation. Among them we mention the stochastic version of the Verhulst model, the stochastic logistic models that are counterparts of the deterministic logistic models treated by Tsoularis and Wallace [72], the SIR model with demography that has been used to study the spread of childhood infections, and the stochastic version of the Ross malaria model. Some early results for the Verhulst model are given by Nåsell [47], for the SIR model with demography by Nåsell [46, 50], and for the Ross malaria model by Nåsell [42].

2.3 A Contact Process on a Complete Graph

The assumptions about homogeneity and about uniform exposure made in the model formulation in the previous section are clearly unrealistic. Additional realism can be approached by allowing for spatial influences in the following way. We imagine that a susceptible individual y can become infected by an infective individual x only if y is a neighbor of x. We use knowledge about neighbors to construct a connected graph where the vertex x is connected to all its neighbors y by an edge. Furthermore, a vertex is empty if the corresponding individual is susceptible, while it is occupied by one individual if that individual is infective. Infection of a target site is possible only if the corresponding individual is susceptible. The infection rate is proportional to the proportion of neighbors that are occupied. Recovery of an infective individual is assumed to take place with a constant recovery rate, the same for all occupied sites. This set-up corresponds to the so-called contact process introduced by Harris [32].

A special case of the contact process appears if the connected graph is complete, which means that all its vertices are connected. It is easy to see that the number of sites occupied in the contact process is exactly the number of infected individuals in the SIS process described in the previous section if the connected graph is complete. This connection between the processes is nicely described by Dickman and Vidigal [23].

2.4 Some Early Stochastic Logistic Models

The analysis of the stochastic SIS model formulated in Sect. 2.2 is not elementary. It requires a serious effort, as witnessed by the content of the present monograph. Early work on this or similar logistic models avoided some of the mathematical difficulties by various reformulations. We mention these contributions briefly.

The earliest paper of this type is Kendall [35]. He introduces the restriction that the state space is strictly positive. His process therefore lacks an absorbing state. This simplifies the mathematical treatment considerably, since the difficulty of dealing with an absorbing state is avoided. This means that the two phenomena that are at the heart of our investigation, namely extinction and quasi-stationarity, cannot be studied. The Kendall process is closely related to the auxiliary process $\{X^{(0)}(t)\}$ that is introduced in the next chapter.

Whittle [77] introduces an approximation method that later has been referred to as moment closure. He applies his method to the Kendall model. Good agreement is reached in a specific numerical example. Our findings later on in this monograph indicate that Whittle's approximation method applied to Kendall's approximating process may give reasonable results in one parameter region, but poor results in other regions.

Another variation of the Verhulst logistic model is studied by Prendiville [62]. He proposes that the population birth rate λ_n and the population death rate μ_n, which are quadratic functions of n in the Verhulst logistic model, be replaced by linear functions of n, with the birth rate decreasing and the death rate increasing functions of n. As in the Kendall case, the state space excludes the state zero. Thus, the phenomenon of extinction is absent also from this model. The stationary distribution is non-degenerate and can be determined explicitly. The same model is studied by Takashima [70].

The term "logistic process" is used by Ricciardi [65] to refer to the Prendiville process. Bharucha-Reid [14] gives a brief discussion of what he calls the stochastic analogue of the logistic law of growth. He accounts for the work on the models formulated by Kendall and Prendiville. Iosifescu and Tautu [33] also discuss these early contributions. Kijima [36] also refers to the Prendiville process as a logistic process.

All the authors mentioned in this section analyse stochastic processes that they refer to as "logistic processes", but none of them deals with the process that we are concerned with here. The appropriate stochastic version of any logistic model has an absorbing state at the origin, and its deterministic version is nonlinear.

2.5 Parameter Spaces

It is important for the further analysis to realize that the parameter spaces of the deterministic and stochastic versions of the model are different. The analysis of the parameter spaces is exceedingly simple in the present case, since so few parameters are involved. However, the ideas used are important both in this particular case and in more complicated models. In the discussion we make use of the classification of parameters as "essential" and "innocent", introduced by Nåsell [48]. A parameter is termed innocent if it can be eliminated by a rescaling in the deterministic version of the model of either a state variable or of time. Any variable that is not innocent is called essential.

We note from Sect. 2.2 that three parameters have been introduced to describe the hypotheses of the SIS model, namely the contact rate λ, the recovery rate μ, and the population size N. Furthermore, we have reparametrized by writing $R_0 = \lambda/\mu$. Among the resulting three parameters R_0, N, and μ, we note first that the recovery rate μ is an innocent parameter since it can be eliminated by rescaling time from t to $s = \mu t$. Furthermore, the basic reproduction ratio R_0 can not be eliminated by rescaling. Therefore, it is an essential parameter. Finally, the population size N plays different roles in the deterministic and stochastic versions of the model. It can clearly be eliminated from the deterministic version of the model by defining $i = I/N$ as the proportion of infected individuals, and deriving a differential equation for i. However, the corresponding operation is not meaningful in the stochastic version of the model. Indeed, the stochastic model is a model for finite populations, while the deterministic model is a model for proportions. This is consistent with the known fact that the deterministic version of the model can be derived from the stochastic one by first scaling the state variable by N, and then letting N approach infinity. In this way, the population size N disappears from the scene.

We conclude that the parameter space for the deterministic model contains only one essential parameter, namely R_0, while that for the stochastic model contains two essential parameters, namely R_0 and N.

This difference between the parameter spaces for the two versions of the model has an important consequence for the analysis of the stochastic model. Indeed, we shall derive asymptotic results that are valid for large values of the population size N. In this situation it is important to establish parameter regions in which the quantities studied are qualitatively similar. Here, we take a clue from the threshold result in the deterministic version of the model. It can be used to partition the parameter space into two regions with qualitatively different behavior. The parameter space in this case consists of all positive values of the basic reproduction ratio R_0, and the partition given by the threshold result shows that the two regions where $0 < R_0 < 1$ and $1 < R_0$ give qualitatively different results. We can therefore expect these two regions to give qualitatively different results also in the stochastic case. As already mentioned in the Introduction, we describe the corresponding parameter regions for the stochastic version of the model by saying that R_0 is distinctly above the deterministic threshold value one in the first case, and distinctly below this threshold value in the second case. This means for finite N that $R_0 - 1$ is larger than a positive and decreasing function of N in the first case, and smaller than a positive and decreasing function of N in the second case. The region between these two boundaries identifies the so-called transition region. It is described by introducing the reparametrization

$$\rho = (R_0 - 1)\sqrt{N}. \tag{2.16}$$

The reasoning that leads to it is given in Sect. 7.3. Formally, the transition region is defined by requiring that ρ is constant as $N \to \infty$. This implies that $R_0 \to 1$ as $N \to \infty$ in the transition region. Of course, the rate at which R_0 approaches zero as $N \to \infty$ in the transition region depends on the reparametrization, which is model dependent. We mention the useful rule of thumb that says that R_0 is in the transition region if the absolute value of ρ is smaller than about three.

Chapter 3
Stochastic Process Background

This chapter contains some stochastic process background that is of value for the study of the SIS model. In the first seven sections of the chapter we give results for a much larger class of stochastic processes than the specific SIS model, while the last two sections deal with the SIS model. The results in the first seven sections are expected to be of value in the study of quasi-stationarity for other stochastic logistic models than the SIS model that is the main object of study throughout the monograph.

In Sect. 3.1 we formulate a birth–death process with finite state space, and for which the origin is an absorbing state. The formulation is based on a set of transition rates λ_n and μ_n, referred to as birth rates and death rates, respectively. This process has the important feature that extinction will occur in finite time, and that the distribution of states will for some parameter values remain nearly constant for a long time before it goes extinct. This nearly constant distribution is referred to as the quasi-stationary distribution. It can not be determined explicitly. The main goal of the present monograph is to derive useful approximations. Following Cavender [16] and Kryscio and Lefèvre [39] we study two auxiliary processes without absorbing states in Sect. 3.2. It is straightforward to determine their stationary distributions, called $\mathbf{p}^{(0)}$ and $\mathbf{p}^{(1)}$, in terms of the basic transition rates. It turns out that they play important roles in the search for approximations of the quasi-stationary distribution. The quasi-stationary distribution $\mathbf{q}$ of the original process is studied in Sect. 3.3. It leads us to define a map Ψ between discrete distributions on $\{1, 2, \ldots, N\}$ with the property that $\mathbf{q}$ is the unique fixed point of Ψ. The time to extinction is studied in Sect. 3.4. It turns out to be closely related to the quasi-stationary distribution and to the stationary distribution of one of the two auxiliary processes.

Stochastic ordering results are of considerable importance for our study. Cavender [16] started by proving that the stationary distribution $\mathbf{p}^{(0)}$ provides a lower bound of the quasi-stationary distribution. Kryscio and Lefèvre [39] added to this the conjecture that the other stationary distribution, $\mathbf{p}^{(1)}$, would provide an upper bound of the quasi-stationary distribution. This conjecture remained an open problem until it was solved by Clancy and Pollett [18]. The older proof concerning the lower

I. Nåsell, *Extinction and Quasi-stationarity in the Stochastic Logistic SIS Model*, Lecture Notes in Mathematics 2022, DOI 10.1007/978-3-642-20530-9_3, © Springer-Verlag Berlin Heidelberg 2011

bound was given in the general situation that we deal with in the first seven sections of this chapter, where no particular assumptions are made on the transition rates λ_n and μ_n, while the harder proof that dealt with the upper bound was based on the particular expressions for λ_n and μ_n that define the SIS model, and that are given by (2.14) and (2.15) in Sect. 2.2.

Results concerning ordering of distributions are dealt with in Sects. 3.5–3.8. Two different ordering relations are defined in Sect. 3.5, and one of them is shown to imply the other one. In Sect. 3.6 we establish an important result due to Clancy and Pollett. It shows that the map Ψ mentioned above has the property that it preserves stochastic ordering. Section 3.7 is then used to give an elegant proof, due to Clancy and Pollett, of the result first established by Cavender, namely that the stationary distribution $\mathbf{p}^{(0)}$ provides a lower bound of the quasi-stationary distribution $\mathbf{q}$. The result that the other stationary distribution, $\mathbf{p}^{(1)}$, provides an upper bound of the quasi-stationary distribution $\mathbf{q}$ for the SIS model is established in Sect. 3.8.

3.1   Formulation of a Birth–Death Process

We study a birth–death process $\{X(t), t \geq 0\}$ with finite state space $\{0, 1, \ldots, N\}$ and where the origin is an absorbing state. Transitions in the process are only allowed to neighboring states. The rate of transition from state n to state $n + 1$ (the population birth rate) is denoted λ_n, and the rate of transition from the state n to the state $n - 1$ (the population death rate) is denoted by μ_n. It is convenient to also have notation for the sum of λ_n and μ_n: we put $\kappa_n = \lambda_n + \mu_n$. We assume that μ_0 and λ_N are equal to zero, to be consistent with the assumption that the state space is limited to $\{0, 1, \ldots, N\}$. Furthermore, we assume that λ_0 also equals zero, to be consistent with the assumption that the origin is an absorbing state. All other transition rates are assumed to be strictly positive.

The Kolmogorov forward equations for the state probabilities $p_n(t) = P\{X(t) = n\}$ can be written

$$p_n'(t) = \mu_{n+1}p_{n+1}(t) - \kappa_n p_n(t) + \lambda_{n-1}p_{n-1}(t), \quad n = 0, 1, \ldots, N. \tag{3.1}$$

(Put $\mu_{N+1} = \lambda_{-1} = p_{N+1}(t) = p_{-1}(t) = 0$, so that (3.1) makes sense formally for all n-values indicated.) The state probabilities depend on the initial distribution $p_n(0)$.

An alternative way of writing this system of equations is in the form

$$\mathbf{p}' = \mathbf{p}\mathbf{Q}, \tag{3.2}$$

where $\mathbf{p}(t) = (p_0(t), p_1(t), \ldots, p_N(t))$ is the row vector of state probabilities and the matrix $\mathbf{Q}$ contains the transition rates as follows: The nondiagonal element q_{mn} of this matrix equals the rate of transition from state m to state n, and the diagonal element q_{nn} equals the rate κ_n multiplied by -1. The matrix $\mathbf{Q}$ can be written as follows:

$$Q = \begin{pmatrix} -\kappa_0 & \lambda_0 & 0 & \ldots & 0 \\ \mu_1 & -\kappa_1 & \lambda_1 & \ldots & 0 \\ 0 & \mu_2 & -\kappa_2 & \ldots & 0 \\ \ldots & \ldots & \ldots & \ldots & \ldots \\ 0 & 0 & 0 & \ldots & -\kappa_N \end{pmatrix} \tag{3.3}$$

Note that Q is a tridiagonal matrix with all row sums equal to 0. Thus the determinant of Q equals zero, so Q is noninvertible. Note also that $\lambda_0 = \kappa_0 = 0$, so the first row of Q is a row of zeros.

A stationary distribution is found by putting the time derivative equal to zero, i.e. by solving $pQ = 0$. It is readily shown that the process $\{X(t)\}$ has a degenerate stationary distribution $p = (1, 0, \ldots, 0)$. The distribution of $\{X(t)\}$ approaches the stationary distribution as time t approaches infinity. This says that ultimate absorption is certain. In many cases, the time to absorption is long. It is therefore of interest to study the distribution of $\{X(t)\}$ before absorption has taken place. This is done via the concept of quasi-stationarity, which is introduced in Sect. 3.3.

3.2   Two Auxiliary Processes

In this section we study two birth–death processes $\{X^{(0)}(t)\}$ and $\{X^{(1)}(t)\}$ that both are close to the original process $\{X(t)\}$, but lack absorbing states. The state space of each of the two auxiliary processes coincides with the set of transient states $\{1, 2, \ldots, N\}$ for the original process. We determine the stationary distribution of each of the two auxiliary processes.

The process $\{X^{(0)}(t)\}$ can be described as the original process with the origin removed. Its death rate $\mu_1^{(0)}$ from the state 1 to the state 0 is equal to 0, while all other transition rates are equal to the corresponding rates for the original process.

The process $\{X^{(1)}(t)\}$ is found from the original process by allowing for one permanently infected individual. Here, each death-rate μ_n is replaced by $\mu_n^{(1)} = \mu_{n-1}$, while each of the birth rates $\lambda_n^{(1)}$ equals the corresponding birth rate λ_n for the original process.

The state probabilities for the first auxiliary process are stored in the probability vector $p^{(0)}(t) = (p_1^{(0)}(t), p_2^{(0)}(t), \ldots, p_N^{(0)}(t))$, and those for the second one are in $p^{(1)}(t) = (p_1^{(1)}(t), p_2^{(1)}(t), \ldots, p_N^{(1)}(t))$.

Stationary distributions are easy to determine explicitly for both of the auxiliary processes. In order to describe them we introduce two sequences ρ_n and π_n as follows:

$$\rho_1 = 1, \quad \rho_n = \frac{\lambda_1 \lambda_2 \cdots \lambda_{n-1}}{\mu_1 \mu_2 \cdots \mu_{n-1}}, \quad n = 2, 3, \ldots, N, \tag{3.4}$$

and

$$\pi_1 = 1, \quad \pi_n = \frac{\lambda_1 \lambda_2 \cdots \lambda_{n-1}}{\mu_2 \mu_3 \cdots \mu_n}, \quad n = 2, 3, \ldots, N. \tag{3.5}$$

It is useful to note that

$$\pi_n = \frac{\mu_1}{\mu_n}\rho_n, \quad n = 1, 2, \ldots, N. \tag{3.6}$$

The two stationary distributions can be simply expressed in terms of these sequences. The stationary distribution of the process $\{X^{(0)}(t)\}$ is given by

$$p_n^{(0)} = \pi_n p_1^{(0)}, \quad n = 1, 2, \ldots, N, \quad \text{where} \quad p_1^{(0)} = \frac{1}{\sum_{n=1}^{N} \pi_n}, \tag{3.7}$$

while the stationary distribution of the process $\{X^{(1)}(t)\}$ is

$$p_n^{(1)} = \rho_n p_1^{(1)}, \quad n = 1, 2, \ldots, N, \quad \text{where} \quad p_1^{(1)} = \frac{1}{\sum_{n=1}^{N} \rho_n}. \tag{3.8}$$

Both of these stationary distributions serve as approximations of the quasi-stationary distribution defined in the next section. Furthermore, the related sequences ρ_n and π_n play important roles in a recursion relation for the quasi-stationary distribution, and in an explicit expression for the expected time to extinction from an arbitrary initial distribution.

3.3   The Quasi-stationary Distribution q

We define and study the quasi-stationary distribution $\mathbf{q}$ of the process $\{X(t)\}$ in this section.

We partition the state space into two subsets, one containing the absorbing state 0, and the other equal to the set of transient states $\{1, 2, \ldots, N\}$. Corresponding to this partition, we write the equation (3.2) in block form. The vector $\mathbf{p}(t)$ is expressed as $\mathbf{p}(t) = (p_0(t), \mathbf{p_T}(t))$, where $\mathbf{p_T}(t) = (p_1(t), \ldots, p_N(t))$ is the row vector of state probabilities in the set of transient states. The corresponding block form of the matrix $\mathbf{Q}$ contains four blocks. The first row of $\mathbf{Q}$ gives rise to two blocks of row vectors of zeroes. The remaining two blocks contain a column vector $\mathbf{R}$ of length N and a square matrix $\mathbf{T}$ of order N. The first entry of $\mathbf{R}$ equals μ_1, while all other entries are equal to 0. The matrix $\mathbf{T}$ is formed be deleting the first row and the first column from $\mathbf{Q}$. With this notation one can rewrite (3.2) in the form

$$(p_0'(t), \mathbf{p_T}'(t)) = (p_0(t), \mathbf{p_T}(t)) \begin{pmatrix} 0 & 0 \\ \mathbf{R} & \mathbf{T} \end{pmatrix}. \tag{3.9}$$

By carrying out the product on the right-hand side and equating the block components of the two sides of the equation, we are led to the following two differential equations:

$$p_0'(t) = \mathbf{p_T}(t)\mathbf{R} = \mu_1 p_1(t) \tag{3.10}$$

and

$$\mathbf{p_T}'(t) = \mathbf{p_T}(t)\mathbf{T}. \tag{3.11}$$

Before absorption, the process takes values in the set of transient states. The state of the process at time t is restricted to this set if two conditions are fulfilled. One is that the initial distribution is supported on this set, i.e. that $P\{X(0) > 0\} = 1$. The second condition is that absorption at the origin has not occurred at time t, i.e. that $X(t) > 0$. The corresponding conditional state probabilities are denoted $\tilde{q}_n(t)$. It is also useful to put $\tilde{\mathbf{q}}(t) = (\tilde{q}_1(t), \ldots, \tilde{q}_N(t))$ to denote the row vector of conditional state probabilities. We note that $\tilde{\mathbf{q}}(t)$ depends on the initial distribution $\tilde{\mathbf{q}}(0)$. The conditioning on non-absorption at time t leads to the relation

$$\tilde{q}_n(t) = P\{X(t) = n | X(t) > 0\} = \frac{p_n(t)}{1 - p_0(t)}. \tag{3.12}$$

Hence the vector of conditional state probabilities $\tilde{\mathbf{q}}(t)$ can be determined from the vector $\mathbf{p_T}(t)$ of state probabilities on the set of transient states via the relation

$$\tilde{\mathbf{q}}(t) = \frac{\mathbf{p_T}(t)}{1 - p_0(t)}. \tag{3.13}$$

By differentiating this relation and using expressions (3.10) and (3.12) we get

$$\tilde{\mathbf{q}}'(t) = \frac{\mathbf{p_T}'(t)}{1 - p_0(t)} + \mu_1 \tilde{q}_1(t)\frac{\mathbf{p_T}(t)}{1 - p_0(t)}. \tag{3.14}$$

By using (3.11) and (3.13) we get the following differential equation for the vector of conditional state probabilities $\tilde{\mathbf{q}}$:

$$\tilde{\mathbf{q}}'(t) = \tilde{\mathbf{q}}(t)\mathbf{T} + \mu_1 \tilde{q}_1(t)\tilde{\mathbf{q}}(t). \tag{3.15}$$

The quasi-stationary distribution $\mathbf{q}$ can now be defined. It is the stationary solution of this equation. Thus, it satisfies the equation

$$\mathbf{q}\mathbf{T} = -\mu_1 q_1 \mathbf{q}. \tag{3.16}$$

This shows that the quasi-stationary distribution $\mathbf{q}$ is a left eigenvector of the matrix $\mathbf{T}$ corresponding to the eigenvalue $-\mu_1 q_1$. This result can be used for numerical evaluations. One can show that the eigenvalue $-\mu_1 q_1$ is the maximum eigenvalue of the matrix $\mathbf{T}$.

We show that the probabilities q_n that determine the quasi-stationary distribution satisfy the following relation:

$$q_n = \pi_n \sum_{k=1}^{n} \frac{1 - \sum_{j=1}^{k-1} q_j}{\rho_k} q_1, \quad n = 1, 2 \ldots, N, \quad \sum_{n=1}^{N} q_n = 1. \tag{3.17}$$

This relation is basic for our study of the quasi-stationary distribution. We use it both for numerical evaluations and for derivation of explicit approximations.

Note that the sequences π_n and ρ_n are known explicitly in terms of the transition rates λ_n and μ_n. We recall that they are related to the two stationary distributions of the previous section, since $\pi_n = p_n^{(0)}/p_1^{(0)}$ and $\rho_n = p_n^{(1)}/p_1^{(1)}$.

In order to derive the result in (3.17) we note from relation (3.16) that the probabilities q_n satisfy the following difference equation of order two:

$$\mu_{n+1}q_{n+1} - \kappa_n q_n + \lambda_{n-1}q_{n-1} = -\mu_1 q_1 q_n, \quad n = 1, 2, \ldots, N, \tag{3.18}$$

with boundary conditions given by

$$q_0 = 0 \quad \text{and} \quad q_{N+1} = 0. \tag{3.19}$$

This difference equation of order two is equivalent to two difference equations of order one. We derive them and solve them consecutively. We define

$$f_n = \mu_n q_n - \lambda_{n-1}q_{n-1}, \quad n = 1, 2, \ldots, N+1. \tag{3.20}$$

Then (3.18) can be written

$$f_{n+1} - f_n = -\mu_1 q_1 q_n, \quad n = 1, 2, \ldots, N, \tag{3.21}$$

with the final condition

$$f_{N+1} = 0. \tag{3.22}$$

We solve this difference equation for f_n. Starting with the n-values N and $N-1$, we get

$$f_N = \mu_1 q_1 q_N, \tag{3.23}$$

$$f_{N-1} = f_N + \mu_1 q_1 q_{N-1} = \mu_1 q_1 [q_{N-1} + q_N], \tag{3.24}$$

and therefore

$$f_n = \mu_1 q_1 \sum_{i=n}^{N} q_i, \quad n = 1, 2, \ldots, N+1. \tag{3.25}$$

The second difference equation of order one is found by inserting this expression for f_n into the expression (3.20) that defines f_n. Thus, we get

$$\mu_n q_n = \lambda_{n-1}q_{n-1} + \mu_1 q_1 \sum_{i=n}^{N} q_i, \quad n = 1, 2, \ldots, N+1, \tag{3.26}$$

with the initial condition

$$q_0 = 0. \tag{3.27}$$

We solve this difference equation for q_n. With $n = 1$ we get

$$\mu_1 q_1 = \mu_1 q_1 \sum_{i=1}^{N} q_i, \tag{3.28}$$

which implies

$$\sum_{i=1}^{N} q_i = 1. \tag{3.29}$$

Setting $n = 2$ gives

$$\mu_2 q_2 = \lambda_1 q_1 + \mu_1 q_1 \sum_{i=2}^{N} q_i = \lambda_1 q_1 + \mu_1 q_1 (1 - q_1). \tag{3.30}$$

Solving for q_2, we get

$$q_2 = \frac{\lambda_1}{\mu_2} q_1 + \frac{\mu_1}{\mu_2} q_1 (1 - q_1) = \frac{\lambda_1}{\mu_2} q_1 \left[1 + \frac{\mu_1}{\lambda_1} (1 - q_1) \right]$$

$$= \pi_2 q_1 \left[\frac{1}{\rho_1} + \frac{1 - q_1}{\rho_2} \right] = \pi_2 q_1 \sum_{k=1}^{2} \frac{1 - \sum_{j=1}^{k-1} \rho_j}{\rho_k}. \tag{3.31}$$

By continuing in this way, we find that (3.17) holds.

We emphasize that the relation in (3.17) is not an explicit solution. It can be used to successively determine the values of q_2, q_3 etc. if q_1 is known. But the crux is that q_1 can only be determined from the relation $\sum_{n=1}^{N} q_n = 1$, which requires knowledge of all the q_n.

Two iteration methods for determining the quasi-stationary distribution can be based on (3.17). One of these methods uses iteration for determining the probability q_1. It starts with an initial guess for q_1, determines successively all the q_n from (3.17), computes the sum of the q_n, and determines the result of the first iteration as the initial guess divided by this sum. The process is repeated until successive iterates are sufficiently close. Nisbet and Gurney [52] describe essentially the same iteration method, based on (3.18), and use it for numerical evaluations.

The second method uses iteration on the whole distribution. To describe it, we define a map Ψ between discrete distributions on $\{1, 2, \ldots, N\}$ as follows: Let $v = (v_1, v_2, \ldots, v_N)$ be a given probability vector, and consider the variation of (3.17) that consists in solving for q_n under the condition that the quantities q_j that appear in the numerator in the sum over k are replaced by the known values v_j. The solution is then interpreted as the image under Ψ of the vector v. The map Ψ thus defined serves an important role in our search for approximations of the quasi-stationary distribution. It will reappear with a different interpretation in Sect. 3.6.

We give a formal description of Ψ. We use $\tilde{\mathbf{p}}$ to denote the image of ν under Ψ. Thus we have

$$\tilde{\mathbf{p}} = \Psi(\nu). \tag{3.32}$$

By using (3.17) we find that the components of $\tilde{\mathbf{p}}$ can be determined as follows:

$$\tilde{p}_n = \pi_n S_n \tilde{p}_1, \quad n = 1, 2, \ldots, N, \tag{3.33}$$

where

$$S_n = \sum_{k=1}^{n} r_k, \quad n = 1, 2, \ldots, N, \tag{3.34}$$

and

$$r_k = \frac{1 - \sum_{j=1}^{k-1} v_j}{\rho_k}, \quad k = 1, 2, \ldots, N, \tag{3.35}$$

with

$$\tilde{p}_1 = \frac{1}{\sum_{n=1}^{N} \pi_n S_n}. \tag{3.36}$$

The map Ψ serves to define an iteration scheme as follows: Successive iterates are numbered with a superscript inside brackets. The iteration starts with an initial guess $\mathbf{q}^{[0]}$ for the quasi-stationary distribution; the stationary distributions $\mathbf{p}^{(0)}$ and $\mathbf{p}^{(1)}$ discussed in the previous section are candidates. The first iteration step determines $\mathbf{q}^{[1]}$ by applying Ψ to $\mathbf{q}^{[0]}$, and later iteration steps determine $\mathbf{q}^{[i]} = \Psi(\mathbf{q}^{[i-1]}) = \Psi^i(\mathbf{q}^{[0]})$. In the case when we wish to evaluate the quasi-stationary distribution numerically, we repeat the process until successive iterates are sufficiently close.

There are a few cases where the quasi-stationary distribution can be determined explicitly. Thus, the linear birth–death process determined by $\lambda_n = n\lambda$ and $\mu_n = n\mu$ on the infinite state space $\{0, 1, 2, \ldots\}$ has its quasi-stationary distribution equal to the geometric distribution $q_n = (1 - R_0) R_0^{n-1}$ when $R_0 = \lambda/\mu < 1$. Furthermore, the random walk in continuous time with absorbing barrier at the origin, determined by $\lambda_n = \lambda$ and $\mu_n = \mu$ for $n \geq 1$ and $\lambda_0 = \mu_0 = 0$, has the quasi-stationary distribution $q_n = (1 - \sqrt{R_0})^2 n (\sqrt{R_0})^{n-1}$, $n = 1, 2, \ldots$, again under the assumption that $R_0 < 1$. These results have been derived by a number of authors with different methods; see Seneta [67], Cavender [16], Pollett [59, 60], van Doorn [73]. These solutions can be seen to satisfy (3.17) with $N = \infty$.

These two examples of explicit expressions for quasi-stationary distributions have two limitations. One is that they deal with birth–death processes where the transition rates λ_n and μ_n are at most linear functions of the state n, and another one is that they are confined to a parameter region where the time to extinction is short. The main applied interest in quasi-stationarity is, however, with models that account for density dependence so that one (or both) of the transition rates λ_n and μ_n depends nonlinearly on n. Furthermore, the parameter regions where the time to extinction is very long and moderately long are more interesting than the region where the time to extinction is short.

Another example of a birth–death process with a quasi-stationary distribution that can be determined explicitly is the SIS model with $N = 2$. Here it is easy to show that $p_1^{(1)} = 1/(1+R_0/2)$, $p_2^{(1)} = 1 - p_1^{(1)}$, $p_1^{(0)} = 1/(1+R_0/4)$, $p_2^{(0)} = 1 - p_1^{(0)}$, $q_1 = \left(R_0 + 6 - \sqrt{R_0^2 + 12R_0 + 4}\right)/4$, and $q_2 = 1 - q_1$. Also, the inequalities $p_1^{(1)} < q_1 < p_1^{(0)}$ hold, consistent with the stochastic bounds that will be established later. But this case is just a curiosity. Our main interest is in the SIS model with large values of N, and explicit solutions are then not available.

A fruitful strategy is therefore to search for approximations. We shall derive approximations of the quasi-stationary distribution of the SIS model in a sequence of steps. The first step gives approximations of the two stationary distributions $\mathbf{p}^{(1)}$ and $\mathbf{p}^{(0)}$, and the next one gives approximations of the images under Ψ of these two distributions, and also approximations of the images under repeated application of Ψ of the stationary distribution $\mathbf{p}^{(1)}$. After that we use results to be established later in this chapter on stochastic bounds to derive approximations of the quasi-stationary distribution $\mathbf{q}$. All of these approximations are given separately in each of three parameter regions. The final step determines an approximation of the quasi-stationary distribution that is valid uniformly across all three parameter regions.

Quasi-stationarity is a subtle phenomenon. This has led several authors to refer to the stationary distribution $\mathbf{p}^{(0)}$ of the auxiliary process $\{X^{(0)}(t)\}$ as the quasi-stationary distribution $\mathbf{q}$ of the original one. The reason for this incorrect reference is probably that the original process $\{X(t)\}$ and the auxiliary process $\{X^{(0)}(t)\}$ have the same transition rates as long as the first process has not gone extinct. However, it should be noted that the two processes differ with regard to one quantity, namely the time spent in state one. This time is exponentially distributed in both of the processes, but the expectation of this time differs. For the original process the expected time in this state equals $1/(\lambda_1 + \mu_1)$, while it is larger and equal to $1/\lambda_1$ for the auxiliary process.

3.4   The Time to Extinction

The time to extinction τ is a random variable that clearly depends on the initial distribution. We denote this random variable by τ_Q when the initial distribution $\mathbf{p}(0)$ equals the quasi-stationary distribution $\mathbf{q}$, and by τ_n when $X(0) = n$.

If absorption has occurred at time t, then clearly the waiting time to extinction τ is at most equal to t and also the state of the process $X(t)$ is equal to 0. Hence the events $\{\tau \leq t\}$ and $\{X(t) = 0\}$ are identical. By computing the probabilities of these events we get

$$P\{\tau \leq t\} = P\{X(t) = 0\} = p_0(t). \tag{3.37}$$

We study first τ_Q. The reason why this is interesting is as follows. One can show that the distribution of the process, conditioned on non-extinction, approaches the quasi-stationary distribution as time increases, see van Doorn [73]. This holds for arbitrary initial distributions that are supported on the set of transient states. If the process has been going on for a long time, and it is known that it has not been absorbed, then its distribution is approximated by the quasi-stationary distribution.

It turns out to be possible to get a simple form for the state probabilities **p** in this case. To derive this we note from (3.14) that

$$\mathbf{p_T}'(t) = -\mu_1 q_1 \mathbf{p_T}(t), \quad \mathbf{p_T}(0) = \mathbf{q}. \tag{3.38}$$

This says essentially that probability is leaking from each transient state with the same rate. This equation has the solution

$$\mathbf{p_T}(t) = \mathbf{q}\exp(-\mu_1 q_1 t). \tag{3.39}$$

From this solution one can also determine $p_0(t)$, since (3.10) gives a differential equation for p_0 in terms of p_1, and the initial value is $p_0(0) = 0$. Thus p_0 satisfies the initial value problem

$$p_0'(t) = \mu_1 p_1(t), \quad p_0(0) = 0, \tag{3.40}$$

with solution

$$p_0(t) = 1 - \exp(-\mu_1 q_1 t). \tag{3.41}$$

The expressions in (3.39) and (3.41) combine to give the solution of the Kolmogorov forward equations in this case. It is noteworthy that it is possible to get such a simple expression for this solution. The solution is, however, not explicit, since we have no explicit expression for the important probability q_1.

We conclude, using (3.37), that the time to extinction from the quasi-stationary distribution, τ_Q, has an exponential distribution with the expectation

$$E\tau_Q = \frac{1}{\mu_1 q_1}. \tag{3.42}$$

The distribution of τ_Q is thus completely determined from the probability q_1.

The distribution of the time to extinction τ from an arbitrary initial distribution is more complicated. Some insight into its behavior may be gained by considering its expectation. It is a standard result for birth–death processes that this expectation can be determined explicitly when $X(0) = n$. Expressions for the result are given e.g. by Karlin and Taylor [34], Gardiner [30], Nisbet and Gurney [52], and Renshaw [64]. We show below that the result can be expressed as follows:

$$E\tau_n = \frac{1}{\mu_1} \sum_{k=1}^{n} \frac{1}{\rho_k} \sum_{j=k}^{N} \pi_j = \frac{1}{\mu_1} \sum_{j=1}^{N} \pi_j \sum_{k=1}^{\min(n,j)} \frac{1}{\rho_k}. \tag{3.43}$$

The second expression follows from the first one by changing the order of summation. Note that the two parameter sequences ρ_n and π_n that appear here are related to the stationary distributions $\mathbf{p}^{(1)}$ and $\mathbf{p}^{(0)}$ of the two auxiliary processes $\{X^{(0)}(t)\}$ and $\{X^{(1)}(t)\}$. Approximations of these stationary distributions can hence be used to derive approximations of $E\tau_n$.

By putting $n = 1$ in the second expression, we find that the expected time to extinction from the state 1 can be written as follows:

$$E\tau_1 = \frac{1}{\mu_1} \sum_{j=1}^{N} \pi_j = \frac{1}{\mu_1 p_1^{(0)}}. \tag{3.44}$$

By inserting the expression $\pi_j = p_j^{(0)}/p_1^{(0)}$ from (3.7) into the first expression for $E\tau_n$ in (3.43), we get

$$E\tau_n = E\tau_1 \sum_{k=1}^{n} \frac{1}{\rho_k} \sum_{j=k}^{N} p_j^{(0)}. \tag{3.45}$$

The expected time to extinction from an arbitrary initial distribution $\{p_n(0)\}$ can be derived from the above expression for $E\tau_n$. The result can be written

$$E\tau = \frac{1}{\mu_1} \sum_{j=1}^{N} \pi_j \sum_{k=1}^{j} \frac{1}{\rho_k} \sum_{n=k}^{N} p_n(0). \tag{3.46}$$

This assumes that the initial distribution is supported on the set of transient states, i.e. that $\sum_{n=1}^{N} p_n(0) = 1$.

Both Nisbet and Gurney [52] and Renshaw [64] give exact analytic expressions for $E\tau_n$, although their notation differs from ours. They describe these expressions as cumbersome and claim that they do not allow an intuitive understanding. Our different notation and our interpretation of ρ_n and π_n in terms of the stationary distributions $\mathbf{p}^{(1)}$ and $\mathbf{p}^{(0)}$ improves this situation. Additional improvement will be reached when we have derived approximations of ρ_n and π_n for the SIS model, as is done in Chaps. 6 and 8.

It is useful to note that our expressions for expected times to extinction are dimensionally correct. Each expectation is proportional to $1/\mu_1$, which is a natural time constant for the process.

We proceed to derive the first expression for $E\tau_n$ in (3.43). Given that the present state is n, the process will remain in this state for a time that is exponentially distributed with expectation $1/\kappa_n$. After that it will make a transition to the state $n+1$ with probability λ_n/κ_n, and to the state $n-1$ with probability μ_n/κ_n. The expected time to extinction from the state n can therefore be written as follows:

$$E\tau_n = \frac{1}{\kappa_n} + \frac{\lambda_n}{\kappa_n} E\tau_{n+1} + \frac{\mu_n}{\kappa_n} E\tau_{n-1}, \quad n = 1, 2, \ldots, N. \tag{3.47}$$

This is a difference equation of order two with boundary values

$$E\tau_0 = 0 \quad \text{and} \quad E\tau_N = \frac{1}{\mu_N} + E\tau_{N-1}. \tag{3.48}$$

We multiply (3.47) by κ_n and use the relation $\kappa_n = \lambda_n + \mu_n$. This leads to the equation

$$\mu_n(E\tau_n - E\tau_{n-1}) - \lambda_n(E\tau_{n+1} - E\tau_n) = 1, \quad n = 1, 2, \ldots, N. \tag{3.49}$$

To solve this equation we define

$$g_n = \rho_n(E\tau_n - E\tau_{n-1}), \quad n = 1, 2, \ldots, N. \tag{3.50}$$

It follows then that g_n satisfies a difference equation of order one. It can be written

$$\frac{\mu_n g_n}{\rho_n} - \frac{\lambda_n g_{n+1}}{\rho_{n+1}} = 1, \quad n = 1, 2, \ldots, N-1, \tag{3.51}$$

or

$$g_n - \frac{\rho_n \lambda_n}{\mu_n \rho_{n+1}} g_{n+1} = \frac{\rho_n}{\mu_n}, \quad n = 1, 2, \ldots, N-1. \tag{3.52}$$

By applying (3.4) and (3.6) we find that g_n satisfies the simple difference equation of order one

$$g_n - g_{n+1} = \frac{\pi_n}{\mu_1}, \quad n = 1, 2, \ldots, N-1. \tag{3.53}$$

The final value for this difference equation is found to be

$$g_N = \rho_N(E\tau_N - E\tau_{N-1}) = \frac{\rho_N}{\mu_N} = \frac{\pi_N}{\mu_1}. \tag{3.54}$$

The difference equation for g_n with the given final value has the solution

$$g_n = \frac{1}{\mu_1} \sum_{j=n}^{N} \pi_j, \quad n = 1, 2, \ldots, N. \tag{3.55}$$

By inserting this expression for g_n into the defining relation (3.50) for g_n we arrive at a difference equation of order one for $E\tau_n$. It can be written

$$E\tau_n - E\tau_{n-1} = \frac{1}{\mu_1} \frac{1}{\rho_n} \sum_{j=n}^{N} \pi_j, \quad n = 1, 2, \ldots, N, \tag{3.56}$$

with initial value

$$E\tau_0 = 0. \tag{3.57}$$

It is straightforward to solve this initial value problem for $E\tau_n$ and to show that its solution is given by the first expression for $E\tau_n$ in (3.43).

It remains to derive the expression (3.46) for $E\tau$ for an arbitrary initial distribution supported on the set of transient states. Clearly we have

$$E\tau = \sum_{n=1}^{N} p_n(0) E\tau_n. \tag{3.58}$$

By using the second expression for $E\tau_n$ in (3.43) and interchanging the order of summation between n and j we get

$$E\tau = \frac{1}{\mu_1} \sum_{j=1}^{N} \pi_j \sum_{n=1}^{N} p_n(0) \sum_{k=1}^{\min(n,j)} \frac{1}{\rho_k}. \tag{3.59}$$

The result follows by interchanging the order of summation between n and k.

It is instructive to confirm that the results in (3.42) and (3.43) can be derived from (3.46). The derivation of (3.42) from (3.46) makes use of the relation (3.17) for the probabilities q_n in terms of the sequences π_n and ρ_n.

3.5   Ordering Relations Between Probability Vectors

Ordering relations between probability vectors are important for our study. Four different ordering relations are defined by Kijima [36]. We give two of them in the present section and show how they are related. Thus, we let $v^{(1)}$ and $v^{(2)}$ denote two probability vectors that describe distributions on $\{1, 2, \ldots, N\}$. Then a so-called *likelihood ratio ordering* is defined if the following condition holds:

$$v_i^{(1)} v_j^{(2)} \geq v_j^{(1)} v_i^{(2)}, \quad 1 \leq i \leq j \leq N. \tag{3.60}$$

If this is so, then we write

$$v^{(1)} \prec_{LR} v^{(2)}, \tag{3.61}$$

and we describe this relation by saying that $v^{(1)}$ is smaller than $v^{(2)}$ in the sense of likelihood ratio ordering.

A second ordering relation is defined by making reference to distribution functions. $v^{(1)}$ is said to be smaller than $v^{(2)}$ in the sense of *stochastic ordering* if the distribution function of the first one lies above the distribution function of the second one, or, equivalently, if the first distribution function lies to the left of the second one. This can be written

$$\sum_{j=1}^{k} v_j^{(1)} \geq \sum_{j=1}^{k} v_j^{(2)}, \quad 1 \leq k \leq N. \tag{3.62}$$

The stochastic ordering relation is denoted by

$$\mathbf{v}^{(1)} \prec_{ST} \mathbf{v}^{(2)}. \tag{3.63}$$

An important relation between these two ordering relations is given in the following theorem:

Theorem 3.1 *Likelihood ratio ordering implies stochastic ordering:*

$$\mathbf{v}^{(1)} \prec_{LR} \mathbf{v}^{(2)} \Rightarrow \mathbf{v}^{(1)} \prec_{ST} \mathbf{v}^{(2)}, \tag{3.64}$$

Proof. Our derivation is inspired by Schmitz [66]. A similar derivation is given by Kijima and Seneta [37]. Thus, we assume that

$$\mathbf{v}^{(1)} \prec_{LR} \mathbf{v}^{(2)}, \tag{3.65}$$

so that (3.60) holds. Letting i and j be fixed for the moment, we find that the following inequalities hold:

$$v_m^{(1)} v_j^{(2)} \geq v_j^{(1)} v_m^{(2)}, \quad 1 \leq m \leq i, \quad 1 \leq i \leq j \leq N. \tag{3.66}$$

Summations over m from 1 to i give

$$\sum_{m=1}^{i} v_m^{(1)} v_j^{(2)} \geq v_j^{(1)} \sum_{n=1}^{i} v_n^{(2)}, \quad 1 \leq i \leq j \leq N. \tag{3.67}$$

This means that

$$\sum_{m=1}^{i} v_m^{(1)} v_k^{(2)} \geq v_k^{(1)} \sum_{n=1}^{i} v_n^{(2)}, \quad i \leq k \leq j, \quad 1 \leq i \leq j \leq N. \tag{3.68}$$

Summations over k from $i+1$ to j give

$$\sum_{m=1}^{i} v_m^{(1)} \sum_{n=i+1}^{j} v_n^{(2)} \geq \sum_{m=i+1}^{j} v_m^{(1)} \sum_{n=1}^{i} v_n^{(2)}, \quad 1 \leq i < j \leq N. \tag{3.69}$$

By adding $\sum_{m=1}^{i} v_m^{(1)} \sum_{n=1}^{i} v_n^{(2)}$ to both sides of this inequality we get

$$\sum_{m=1}^{i} v_m^{(1)} \sum_{n=1}^{j} v_n^{(2)} \geq \sum_{m=1}^{j} v_m^{(1)} \sum_{n=1}^{i} v_n^{(2)}, \quad 1 \leq i < j \leq N. \tag{3.70}$$

By now putting $j = N$ we get

$$\sum_{m=1}^{i} v_m^{(1)} \geq \sum_{n=1}^{i} v_n^{(2)}, \quad 1 \leq i < N. \tag{3.71}$$

Note that equality holds for $i = N$, since then both the left-hand side and the right-hand side equal one. This proves that $v^{(1)}$ is stochastically smaller than $v^{(2)}$, and completes the proof that likelihood ratio ordering implies stochastic ordering.   □

The converse of this theorem does not hold. To show this, we put $v^{(1)} = (1/2, 1/2, 0)$ and $v^{(2)} = (1/2, 0, 1/2)$. Then we get $v^{(1)} \prec_{ST} v^{(2)}$, while the condition that defines likelihood ratio ordering is not satisfied.

In what follows, we shall derive several results that show that one distribution is smaller than a second one in the sense of likelihood ratio ordering. It is helpful in the interpretation of these results to recall that the first distribution is then also smaller than the second one in the sense of stochastic ordering.

3.6   Three Basic Properties of the Map Ψ

We use the present section to show three basic properties of the map Ψ defined by (3.32)–(3.36). These properties are:

1. Ψ is identical to a certain map defined by Ferrari et al. [27].
2. The iteration scheme for determining the quasi-stationary distribution $\mathbf{q}$ described in Sect. 3.3 converges.
3. The map Ψ preserves likelihood ratio ordering.

In Ferrari et al. [27], the authors study quasi-stationary distributions for Markov chains. They define a map $\tilde{\Psi}$ as follows: Let a probability vector $v = (v_1, v_2, \ldots, v_N)$ be given. Define a process related to the finite state birth–death process $X(t)$ defined in Sect. 3.1 by the requirement that whenever the original process reaches the state zero, then it is immediately restarted in some state j with probability v_j, where $1 \leq j \leq N$. The process restarted in this way has the state space $\{1, 2, \ldots, N\}$, which is finite and irreducible. Therefore, the restarted process has a unique stationary distribution $\tilde{\mathbf{p}} = (\tilde{p}_1, \tilde{p}_2, \ldots, \tilde{p}_N)$. A map $\tilde{\Psi}$ is then defined by $\tilde{\Psi}(v) = \tilde{\mathbf{p}}$. Since $\mathbf{q}$ is a quasi-stationary distribution, we have $\tilde{\Psi}(\mathbf{q}) = \mathbf{q}$. Indeed, $\mathbf{q}$ is the unique fixed point of the map $\tilde{\Psi}$.

We show that the map $\tilde{\Psi}$ defined in this way is identical to the map Ψ defined by (3.32)–(3.36). The Kolmogorov forward equation for the restarted process can be written

$$p_n'(t) = \mu_1 v_n p_1(t) + \lambda_{n-1} p_{n-1}(t) - \kappa_n p_n(t) + \mu_{n+1} p_{n+1}(t), \quad n = 1, 2, \ldots, N. \tag{3.72}$$

Thus, the stationary distribution $\tilde{\mathbf{p}}$ satisfies

$$\mu_{n+1}\tilde{p}_{n+1} - \kappa_n\tilde{p}_n + \lambda_{n-1}\tilde{p}_{n-1} = -\mu_1\tilde{p}_1\nu_n, \quad n = 1,2,\ldots,N. \tag{3.73}$$

This equation resembles (3.18) and can be solved in the same way. Thus, we get

$$\tilde{p}_n = \pi_n \sum_{k=1}^{n} \frac{c_k}{\rho_k}\tilde{p}_1, \quad n = 1,2,\ldots,N, \tag{3.74}$$

where $\tilde{p}_1$ is determined so that $\sum_{n=1}^{N}\tilde{p}_n = 1$, and where

$$c_k = \sum_{s=k}^{N} \nu_s. \tag{3.75}$$

Thus, the map $\tilde{\Psi}$ introduced by Ferrari et al. [27] is identical to the map Ψ defined by (3.32)–(3.36). In what follows, we denote the map by Ψ. We note that Ferrari et al. used Φ to refer to this map. We do not follow this usage, since we shall use Φ to denote the normal distribution function.

To show the second property of the map Ψ, namely that the iteration scheme for determining the quasi-stationary distribution $\mathbf{q}$ described in Sect. 3.3 converges, we apply a result derived by Ferrari et al. [27], namely that the sequence $\Psi^n(\mathbf{v})$ converges to the quasi-stationary distribution $\mathbf{q}$ for arbitrary initial distributions $\mathbf{v}$ supported on $\{1,2,\ldots,N\}$.

The third property of the map Ψ is formulated as a theorem:

Theorem 3.2 *The map Ψ preserves likelihood ratio ordering between probability vectors. Thus,*

$$\mathbf{v}^{(1)} \prec_{LR} \mathbf{v}^{(2)} \Rightarrow \Psi\left(\mathbf{v}^{(1)}\right) \prec_{LR} \Psi\left(\mathbf{v}^{(2)}\right). \tag{3.76}$$

Proof. The proof follows Clancy and Pollett [18] closely. Thus, let $\mathbf{v}^{(1)}$ and $\mathbf{v}^{(2)}$ be two distributions on $\{1,2,\ldots,N\}$ for which we assume that

$$\mathbf{v}^{(1)} \prec_{LR} \mathbf{v}^{(2)}. \tag{3.77}$$

We apply the map Ψ to each of these two distributions, and write $\tilde{\mathbf{p}}^{(1)} = \Psi(\mathbf{v}^{(1)})$ and $\tilde{\mathbf{p}}^{(2)} = \Psi(\mathbf{v}^{(2)})$. The components of the two vectors $\tilde{\mathbf{p}}^{(1)}$ and $\tilde{\mathbf{p}}^{(2)}$ can then be written down explicitly. By using (3.74) we get

$$\tilde{p}_n^{(1)} = \pi_n \sum_{k=1}^{n} \frac{c_k^{(1)}}{\rho_k}\tilde{p}_1^{(1)}, \quad n = 1,2,\ldots,N, \tag{3.78}$$

$$\tilde{p}_n^{(2)} = \pi_n \sum_{l=1}^{n} \frac{c_l^{(2)}}{\rho_l}\tilde{p}_1^{(2)}, \quad n = 1,2,\ldots,N, \tag{3.79}$$

where

$$c_k^{(1)} = \sum_{r=k}^{N} v_r^{(1)}, \quad k = 1, 2, \ldots, N, \tag{3.80}$$

$$c_l^{(2)} = \sum_{s=l}^{N} v_s^{(2)}, \quad k = 1, 2, \ldots, N. \tag{3.81}$$

To show that $\tilde{\mathbf{p}}^{(1)} \prec_{LR} \tilde{\mathbf{p}}^{(2)}$, we take $1 \leq i \leq j \leq N$ and consider the following expression:

$$A_{ij} = \frac{\tilde{p}_i^{(1)} \tilde{p}_j^{(2)} - \tilde{p}_j^{(1)} \tilde{p}_i^{(2)}}{\tilde{p}_1^{(1)} \tilde{p}_1^{(2)}} = \pi_i \pi_j \left[\sum_{k=1}^{i} \frac{c_k^{(1)}}{\rho_k} \sum_{l=1}^{j} \frac{c_l^{(2)}}{\rho_l} - \sum_{k=1}^{j} \frac{c_k^{(1)}}{\rho_k} \sum_{l=1}^{i} \frac{c_l^{(2)}}{\rho_l} \right]. \tag{3.82}$$

Clearly, we wish to prove that $A_{ij} \geq 0$. Now, since $i \leq j$, we can write

$$\sum_{l=1}^{j} \frac{c_l^{(2)}}{\rho_l} = \sum_{l=1}^{i} \frac{c_l^{(2)}}{\rho_l} + \sum_{l=i+1}^{j} \frac{c_l^{(2)}}{\rho_l}, \tag{3.83}$$

$$\sum_{k=1}^{j} \frac{c_k^{(1)}}{\rho_k} = \sum_{k=1}^{i} \frac{c_k^{(1)}}{\rho_k} + \sum_{k=i+1}^{j} \frac{c_k^{(1)}}{\rho_k}. \tag{3.84}$$

Inserting these expressions into the right-hand side of (3.82) leads, after cancellation, to the following expression for A_{ij}:

$$A_{ij} = \pi_i \pi_j \left[\sum_{k=1}^{i} \sum_{l=i+1}^{j} \frac{c_k^{(1)} c_l^{(2)}}{\rho_k \rho_l} - \sum_{k=i+1}^{j} \sum_{l=1}^{i} \frac{c_k^{(1)} c_l^{(2)}}{\rho_k \rho_l} \right]. \tag{3.85}$$

By letting the summation variables in the second of the two double sums exchange names, we are led to the following expression for A_{ij}:

$$A_{ij} = \pi_i \pi_j \sum_{k=1}^{i} \sum_{l=i+1}^{j} \frac{B_{kl}}{\rho_k \rho_l}, \tag{3.86}$$

where

$$B_{kl} = c_k^{(1)} c_l^{(2)} - c_l^{(1)} c_k^{(2)}. \tag{3.87}$$

In the double sum in the expression (3.86) for A_{ij} we have $k \leq i < i+1 \leq l$, and therefore $k < l$. With $k < l$ we use the definitions of $c_k^{(1)}$ and $c_l^{(2)}$ in (3.80) and (3.81) to get

$$B_{kl} = \sum_{r=k}^{N} \sum_{s=l}^{N} v_r^{(1)} v_s^{(2)} - \sum_{r=l}^{N} \sum_{s=k}^{N} v_r^{(1)} v_s^{(2)}. \tag{3.88}$$

Now, since $k < l$, we write

$$\sum_{r=k}^{N} v_r^{(1)} = \sum_{r=k}^{l-1} v_r^{(1)} + \sum_{r=l}^{N} v_r^{(1)}, \tag{3.89}$$

$$\sum_{s=k}^{N} v_s^{(2)} = \sum_{s=k}^{l-1} v_s^{(2)} + \sum_{s=l}^{N} v_s^{(2)}. \tag{3.90}$$

Insertion of these sums into the above expression for B_{kl} gives, after cancellation

$$B_{kl} = \sum_{r=k}^{l-1} \sum_{s=l}^{N} v_r^{(1)} v_s^{(2)} - \sum_{r=l}^{N} \sum_{s=k}^{l-1} v_r^{(1)} v_s^{(2)}. \tag{3.91}$$

By changing names of the summation variables in the second of the two double sums, we are led to the following expression for B_{kl}:

$$B_{kl} = \sum_{r=k}^{l-1} \sum_{s=l}^{N} (v_r^{(1)} v_s^{(2)} - v_s^{(1)} v_r^{(2)}). \tag{3.92}$$

Here we use the hypothesis that $v^{(1)} \prec_{LR} v^{(2)}$. In the double sum we have $r < s$, since $r \le l-1 < l \le s$. By using the definition of likelihood ratio ordering we find that all the terms that make up B_{kl} are nonnegative. We conclude that $B_{kl} \ge 0$, and that therefore $A_{ij} \ge 0$. This concludes the proof that Ψ preserves likelihood ratio ordering.   □

The result in this theorem was established by Clancy and Pollett [18] as part of their proof that the stationary distribution $\mathbf{p}^{(1)}$ is a stochastic upper bound of the quasi-stationary distribution $\mathbf{q}$ for the SIS model. This latter property was conjectured to hold true for the SIS model by Kryscio and Lefèvre [39].

3.7 The Stationary Distribution $\mathbf{p}^{(0)}$ Is a Lower Bound of the Quasi-stationary Distribution q

The stationary distributions $\mathbf{p}^{(0)}$ and $\mathbf{p}^{(1)}$ of the two auxiliary processes defined in Sect. 3.2 were originally introduced to provide approximations of the quasi-stationary distribution $\mathbf{q}$, as already mentioned. It is important for our further development to make use of the additional facts that they provide lower and upper bounds, respectively, of the quasi-stationary distribution $\mathbf{q}$. We shall establish that $\mathbf{p}^{(0)}$ provides a lower bound in this section, and defer the treatment of $\mathbf{p}^{(1)}$ as an upper bound to the next section.

The main result in the present section is given by the following theorem:

Theorem 3.3 *The stationary distribution* $\mathbf{p}^{(0)}$ *provides a lower bound of the quasi-stationary distribution* $\mathbf{q}$ *for the finite-state birth–death process with an absorbing state at the origin defined in Sect. 3.1, in the sense of likelihood ratio ordering. Thus,*

$$\mathbf{p}^{(0)} \prec_{LR} \mathbf{q}. \tag{3.93}$$

Proof. Let $\mathbf{e_i}$ be the probability vector that assigns unit mass to state i. Thus, $\mathbf{e_i}$ is a vector with N components, where component number i is equal to one, and all other components are equal to zero. It is then easy to see that $\mathbf{e_1}$ is a lower bound and $\mathbf{e_N}$ is an upper bound of any distribution $\mathbf{p}$ on $\{1, 2, \ldots, N\}$, in the sense of likelihood ratio ordering. Thus, we have

$$\mathbf{e_1} \prec_{LR} \mathbf{p} \prec_{LR} \mathbf{e_N}. \tag{3.94}$$

Since the map Ψ preserves likelihood ratio ordering, as proved in Theorem 3.2, we get

$$\Psi(\mathbf{e_1}) \prec_{LR} \Psi(\mathbf{q}) \prec_{LR} \Psi(\mathbf{e_N}). \tag{3.95}$$

But since $\Psi(\mathbf{q}) = \mathbf{q}$, we conclude from this result that the images under the map Ψ of the vectors $\mathbf{e_1}$ and $\mathbf{e_N}$ supply lower and upper bounds of the quasi-stationary distribution $\mathbf{q}$ in the sense of likelihood ratio ordering:

$$\Psi(\mathbf{e_1}) \prec_{LR} \mathbf{q} \prec_{LR} \Psi(\mathbf{e_N}). \tag{3.96}$$

To prove the claim made in (3.93), namely that the stationary distribution $\mathbf{p}^{(0)}$ provides a lower bound of the quasi-stationary distribution $\mathbf{q}$ in the sense of likelihood ratio ordering, it remains to show that $\Psi(\mathbf{e_1}) = \mathbf{p}^{(0)}$. Indeed, the restarted process with $\mathbf{v} = \mathbf{e_1}$ is a birth–death process where every time state zero is reached, the process is started anew with probability one at the state one. The restarted process is identical to the auxiliary process $X^{(0)}$. Thus, we conclude that Ψ maps the probability vector $\mathbf{e_1}$ onto the stationary distribution $\mathbf{p}^{(0)}$. $\qquad\square$

The elegant proof of this theorem is due to Clancy and Pollett [18].

We note that the proof also shows that $\Psi(\mathbf{e_N})$ provides an upper bound of $\mathbf{q}$ in the sense of likelihood ratio ordering. However, we do not pursue this, since a better upper bound for the SIS model is given in the next section.

A consequence of the above theorem, namely that $\mathbf{p}^{(0)}$ is a lower bound of $\mathbf{q}$ in the sense of stochastic ordering, was established by Cavender [16].

3.8 The Stationary Distribution $\mathbf{p}^{(1)}$ is an Upper Bound of the Quasi-stationary Distribution q for the SIS Model

The story about an upper bound for the quasi-stationary distribution $\mathbf{q}$ of the birth–death process that we are dealing with in this chapter goes back to Kryscio and Lefèvre [39]. They studied the SIS model, and made the important conjecture that the stationary distribution $\mathbf{p}^{(1)}$ provides an upper bound of the quasi-stationary distribution $\mathbf{q}$ in the sense of stochastic ordering for this particular model. This conjecture remained an open problem until it was proved by Clancy and Pollett [18]. Their proof makes use of the important result that they established in the same 2003 paper, namely that the map Ψ defined in Sect. 3.3 preserves likelihood ratio ordering, as shown in Theorem 3.2.

We use the approach taken by Clancy and Pollett to prove the following theorem:

Theorem 3.4 *The stationary distribution* $\mathbf{p}^{(1)}$ *for the SIS model provides an upper bound of the quasi-stationary distribution* $\mathbf{q}$ *in the sense of likelihood ratio ordering, i.e.*

$$\mathbf{q} \prec_{LR} \mathbf{p}^{(1)}. \tag{3.97}$$

Proof. We recall from the discussion in Sect. 3.6 that the map Ψ defined there has the property that the components of the vector $\tilde{\mathbf{p}}$ defined by $\tilde{\mathbf{p}} = \Psi(\mathbf{v})$ can be given explicitly as a function of $\mathbf{v}$. The formula for this is given in (3.74). Another pleasant property of the map Ψ is that it can easily be inverted. Indeed, given any probability vector $\tilde{\mathbf{p}}$, the $\mathbf{v}$-vector components $v_1, v_2, \ldots, v_N$ such that $\Psi(\mathbf{v}) = \tilde{\mathbf{p}}$ can be immediately solved from (3.73). In particular, this means that one can solve the equation $\Psi(\mathbf{v}) = \mathbf{p}^{(1)}$ for $\mathbf{v}$. To do this, we use (3.73) to get

$$v_n = \frac{\lambda_n p_n^{(1)} - \lambda_{n-1} p_{n-1}^{(1)} + \mu_n p_n^{(1)} - \mu_{n+1} p_{n+1}^{(1)}}{\mu_1 p_1^{(1)}}, \quad n = 1, 2, \ldots, N. \tag{3.98}$$

Now the components of the stationary distribution $\mathbf{p}^{(1)}$ satisfy a relation that is similar to the expression in the numerator of v_n. It can be written

$$\lambda_n^{(1)} p_n^{(1)} - \lambda_{n-1}^{(1)} p_{n-1}^{(1)} = \mu_{n+1}^{(1)} p_{n+1}^{(1)} - \mu_n^{(1)} p_n^{(1)}, \quad n = 1, 2, \ldots, N. \tag{3.99}$$

Here we use the relations between the transition rates for the auxiliary process $\{X^{(1)}(t)\}$ and the basic process $\{X(t)\}$. They can be written

$$\lambda_n^{(1)} = \lambda_n, \quad \mu_n^{(1)} = \mu_{n-1}, \quad n = 1, 2, \ldots, N. \tag{3.100}$$

Insertions first into (3.99), and then into (3.98) give

$$v_n = \frac{\mu_n p_{n+1}^{(1)} - \mu_{n-1} p_n^{(1)} + \mu_n p_n^{(1)} - \mu_{n+1} p_{n+1}^{(1)}}{\mu_1 p_1^{(1)}} = \frac{\mu p_n^{(1)} - \mu p_{n+1}^{(1)}}{\mu p_1^{(1)}}$$

$$= \frac{p_n^{(1)} - p_{n+1}^{(1)}}{p_1^{(1)}} = \rho_n - \rho_{n+1}, \quad n = 1, 2, \ldots, N, \quad (3.101)$$

where we have $\rho_{N+1} = 0$. These vector components are nonnegative if $\rho_n \geq \rho_{n+1}$, or, equivalently, $\lambda_n / \mu_n \leq 1$. We note that the sum of the components v_n equals $\sum_{n=1}^N v_n = \rho_1 = 1$.

The next step taken in the proof that $\mathbf{p}^{(1)}$ is an upper bound of q in the sense of likelihood ratio consists in showing that the particular probability vector v that solves $\Psi(v) = \mathbf{p}^{(1)}$ is an upper bound of $\mathbf{p}^{(1)}$. To prove this, we take $1 \leq i \leq j \leq N$, and let the components of v be given by (3.101), and the components of $\mathbf{p}^{(1)}$ by (3.8). We get then

$$p_i^{(1)} v_j - p_j^{(1)} v_i = [\rho_{i+1} \rho_j - \rho_i \rho_{j+1}] p_1^{(1)}, \quad 1 \leq i \leq j \leq N. \quad (3.102)$$

By using the definition of ρ_n in (3.4) we get

$$\rho_{i+1} \rho_j - \rho_i \rho_{j+1} = \rho_i \rho_j \left[\frac{\lambda_i}{\mu_i} - \frac{\lambda_j}{\mu_j} \right], \quad 1 \leq i \leq j \leq N. \quad (3.103)$$

This expression is clearly nonnegative, since $\lambda_i / \mu_i = R_0(1 - i/N)$ is a decreasing function of i for the SIS model. It follows therefore from (3.102) that the stationary distribution $\mathbf{p}^{(1)}$ is a lower bound of v, in the sense of likelihood ratio, or $\mathbf{p}^{(1)} \prec_{LR} v$. Thus, by applying the map Ψ to the probability vector v we have found a lower bound of v, in the sense of likelihood ratio.

At this point we make use of the result established in Theorem 3.2, namely that the map Ψ preserves likelihood ratio ordering. By applying Ψ repeatedly to v, $\Psi(v)$, etc., we get

$$\Psi^i(v) \prec_{LR} \Psi^{i-1}(v) \prec_{LR} \cdots \prec_{LR} \Psi(v) = \mathbf{p}^{(1)} \prec_{LR} v. \quad (3.104)$$

On letting $i \to \infty$, we obtain $\Psi^i(v) \to q$; this is true for any finite-state absorbing chain, as shown by Ferrari et al. [27]. We conclude from this that the stationary distribution $\mathbf{p}^{(1)}$ is an upper bound of the quasi-stationary distribution $\mathbf{q}$, in the sense of likelihood ratio ordering.

We note that if $\lambda_n / \mu_n > 1$ for any n, then the vector v whose components are given by (3.101) has one or more components negative. This means that v can not be interpreted as a probability vector. But the result in Theorem 3.2 that the map Ψ preserves likelihood ratio ordering is still valid for such vectors.

This concludes the proof that the stationary distribution $\mathbf{p}^{(1)}$ provides an upper bound of the quasi-stationary distribution $\mathbf{q}$, in the sense of likelihood ratio ordering, for the SIS model.   □

We note that the previous section and the present one have shown that the stationary distributions $\mathbf{p}^{(0)}$ and $\mathbf{p}^{(1)}$ provide lower and upper bounds of the quasi-stationary distribution $\mathbf{q}$ in the sense of likelihood ratio ordering, for the SIS model. Thus, we have

$$\mathbf{p}^{(0)} \prec_{LR} \mathbf{q} \prec_{LR} \mathbf{p}^{(1)}. \tag{3.105}$$

Improved lower and upper bounds are found by considering the images under the map Ψ of the stationary distributions $\mathbf{p}^{(0)}$ and $\mathbf{p}^{(1)}$. This follows from the beautiful result that Ψ preserves likelihood ratio ordering, and since the quasi-stationary distribution $\mathbf{q}$ is a fixed point of Ψ. Furthermore, it is easy to see that $\mathbf{p}^{(0)} \prec_{LR} \Psi\left(\mathbf{p}^{(0)}\right)$. This follows from the results $\mathbf{e_1} \prec_{LR} \mathbf{p}^{(0)}$ and $\Psi(\mathbf{e_1}) = \mathbf{p}^{(0)}$, and the fact that Ψ preserves likelihood ratio ordering. We summarize these considerations in the following way:

$$\mathbf{p}^{(0)} \prec_{LR} \Psi\left(\mathbf{p}^{(0)}\right) \prec_{LR} \mathbf{q} \prec_{LR} \Psi\left(\mathbf{p}^{(1)}\right) \prec_{LR} \mathbf{p}^{(1)}. \tag{3.106}$$

This relation is of importance in the derivation of approximations of the quasi-stationary distribution $\mathbf{q}$ for the SIS model in Chap. 11.

3.9   Monotonicity of Two Expressions for the SIS Model

This section is used to study two particular expressions that arise when one studies the images under the map Ψ of the two probability vectors $\mathbf{p}^{(1)}$ and $\mathbf{p}^{(0)}$. We shall show that these expressions are monotone functions. These monotonicities will be used repeatedly in Chap. 10.

We introduce $\tilde{\mathbf{p}}^{(1)}$ and $\tilde{\mathbf{p}}^{(0)}$, respectively, to denote the images under Ψ of $\mathbf{p}^{(1)}$ and $\mathbf{p}^{(0)}$. Thus, we have

$$\tilde{\mathbf{p}}^{(1)} = \Psi\left(\mathbf{p}^{(1)}\right), \tag{3.107}$$

$$\tilde{\mathbf{p}}^{(0)} = \Psi\left(\mathbf{p}^{(0)}\right). \tag{3.108}$$

By using the expression (3.32)–(3.36) for the components of the image under the map Ψ of the vector v, we find that

$$\tilde{p}_n^{(1)} = \pi_n \sum_{k=1}^{n} r_k^{(1)} \tilde{p}_1^{(1)}, \quad n = 1, 2, \ldots, N, \tag{3.109}$$

$$\tilde{p}_n^{(0)} = \pi_n \sum_{k=1}^{n} r_k^{(0)} \tilde{p}_1^{(0)}, \quad n = 1, 2, \ldots, N, \tag{3.110}$$

where the quantities $r_k^{(1)}$ and $r_k^{(0)}$ are ratios defined as follows:

$$r_k^{(1)} = \frac{1 - \sum_{j=1}^{k-1} p_j^{(1)}}{\rho_k}, \quad k = 1, 2, \ldots, N, \tag{3.111}$$

$$r_k^{(0)} = \frac{1 - \sum_{j=1}^{k-1} p_j^{(0)}}{\rho_k}, \quad k = 1, 2, \ldots, N. \tag{3.112}$$

We show that these two ratios are decreasing functions of k for the SIS model. As mentioned above, we shall use this result in Chap. 10, where we study approximations of some images under Ψ.

To show that $r_k^{(1)}$ decreases in k, we note first that

$$r_k^{(1)} = \frac{\sum_{j=k}^{N} p_j^{(1)}}{\rho_k} = \frac{\sum_{j=k}^{N} \rho_j}{\rho_k} p_1^{(1)}, \quad k = 1, 2, \ldots, N. \tag{3.113}$$

By using the definition of ρ_k in (3.4) we get

$$\frac{\rho_{k+1}}{\rho_k} = \frac{\lambda_k}{\mu_k}, \quad k = 1, 2, \ldots, N. \tag{3.114}$$

This is decreasing in k for the SIS model. Now the ratio of ratios ρ_{k+i}/ρ_k can be written as a product of i factors of this form, and is therefore decreasing in k. This implies in particular that

$$\frac{\rho_{k+i}}{\rho_k} > \frac{\rho_{k+i+1}}{\rho_{k+1}}, \quad k = 1, 2, \ldots, N - 2, \quad i = 1, 2, \ldots, N - k - 1. \tag{3.115}$$

This decrease of the ratios ρ_{k+i}/ρ_k can be used to show that the ratio $r_k^{(1)}$ is a decreasing function of k. We get

$$\frac{r_k^{(1)} - r_{k+1}^{(1)}}{p_1^{(1)}} = \frac{\sum_{j=k}^{N} \rho_j}{\rho_k} - \frac{\sum_{j=k+1}^{N} \rho_j}{\rho_{k+1}} = \frac{\sum_{i=0}^{N-k} \rho_{k+i}}{\rho_k} - \frac{\sum_{i=0}^{N-k-1} \rho_{k+1+i}}{\rho_{k+1}}$$

$$= \sum_{i=0}^{N-k-1} \left(\frac{\rho_{k+i}}{\rho_k} - \frac{\rho_{k+1+i}}{\rho_{k+1}} \right) + \frac{\rho_N}{\rho_k} > 0, \quad k = 1, 2, \ldots, N - 1. \tag{3.116}$$

The inequality follows since the first term in the last sum is equal to zero, while by (3.115) all remaining terms are positive, as is the single term ρ_N/ρ_k. This completes the proof that the ratios $r_k^{(1)}$ decrease with k.

The proof that $r_k^{(0)}$ also decreases in k is similar. We get

$$r_k^{(1)} = \frac{\sum_{j=k}^N p_j^{(0)}}{\rho_k} = \frac{\sum_{j=k}^N \pi_j}{\rho_k} p_1^{(1)} = \frac{1}{\rho_k} \sum_{j=k}^N \frac{1}{j} \rho_j p_1^{(0)}, \quad k = 1, 2, \ldots, N. \quad (3.117)$$

Hence

$$\frac{r_k^{(0)} - r_{k+1}^{(0)}}{p_1^{(0)}} = \frac{1}{\rho_k} \sum_{j=k}^N \frac{1}{j} \rho_j - \frac{1}{\rho_{k+1}} \sum_{j=k+1}^N \frac{1}{j} \rho_j$$

$$= \frac{1}{\rho_k} \sum_{i=0}^{N-k} \frac{1}{k+i} \rho_{k+i} - \frac{1}{\rho_{k+1}} \sum_{i=0}^{N-k-1} \frac{1}{k+1+i} \rho_{k+1+i}$$

$$= \sum_{i=0}^{N-k-1} \left(\frac{\rho_{k+i}}{(k+i)\rho_k} - \frac{\rho_{k+1+i}}{(k+1+i)\rho_{k+1}} \right) + \frac{\rho_N}{\rho_k}, \quad k = 1, 2, \ldots, N-1. \quad (3.118)$$

Now, by using the inequality in (3.115), we get

$$\frac{\rho_{k+i}}{\rho_k} > \frac{\rho_{k+1+i}}{\rho_{k+1}} > \frac{k+i}{k+1+i} \frac{\rho_{k+1+i}}{\rho_{k+1}}, \quad k = 1, 2, \ldots, N-2, \quad i = 1, 2, \ldots, N-k-1.$$

$$(3.119)$$

Insertion into (3.118) proves that $r_k^{(0)}$ decreases in k.

Chapter 4
The SIS Model: First Approximations
of the Quasi-stationary Distribution

The preceding chapter has given a number of results for a class of stochastic models containing the SIS model. We shall use these results in the chapters that follow, where we gradually derive results that lead up to approximations of the quasi-stationary distribution.

As in Chap. 2, we use $\{I(t)\}$ to denote the birth–death process that serves as the stochastic version of the SIS model. Similarly to Chap. 3, we define two auxiliary processes $\{I^{(0)}(t)\}$ and $\{I^{(1)}(t)\}$ that are slight variations of the original process $\{I(t)\}$. The first one is found from the original process by removing the possibility of transitions from state one to the absorbing state zero, while the second one can be viewed as the original process equipped with one individual who never recovers from infection. The stationary distributions $\mathbf{p}^{(0)}$ and $\mathbf{p}^{(1)}$ of these two auxiliary processes play important roles in our search for approximations of the quasi-stationary distributions. The use of these distributions goes back to the suggestions by Cavender [16] and by Kryscio and Lefèvre [39] that they themselves would serve as approximations of the quasi-stationary distribution.

The main content of the present brief chapter consists in explicit expressions for the two stationary distributions $\mathbf{p}^{(0)}$ and $\mathbf{p}^{(1)}$. They form the starting points for the approximations that are derived in the following chapters. In addition, we give numerical illustrations of the quasi-stationary distribution and of $\mathbf{p}^{(0)}$ and $\mathbf{p}^{(1)}$ for a range of parameter values. This shows that these distributions change character when the basic reproduction ratio R_0 passes the value one. This change in character is a counterpart in the stochastic version of the model to the classical threshold result that holds for the deterministic version of the model. The numerical illustrations also show that the stationary distributions $\mathbf{p}^{(0)}$ and $\mathbf{p}^{(1)}$ do not provide sufficiently good approximations of the quasi-stationary distribution.

The two stationary distributions $\mathbf{p}^{(0)}$ and $\mathbf{p}^{(1)}$ are determined from the two sequences π_n and ρ_n by (3.7) and (3.8). These relations can be expressed as follows:

$$p_n^{(0)} = \frac{\pi_n}{\sum_{n=1}^{N} \pi_n}, \quad n = 1, 2, \dots, N, \tag{4.1}$$

I. Nåsell, *Extinction and Quasi-stationarity in the Stochastic Logistic SIS Model*, Lecture Notes in Mathematics 2022, DOI 10.1007/978-3-642-20530-9_4, © Springer-Verlag Berlin Heidelberg 2011

$$p_n^{(1)} = \frac{\rho_n}{\sum_{n=1}^{N} \rho_n}, \quad n = 1, 2, \ldots, N. \tag{4.2}$$

Furthermore, the two sequences π_n and ρ_n are determined from the transition rates λ_n and μ_n via the definitions in (3.5) and (3.4). Thus, the sequence ρ_n is given by

$$\rho_1 = 1, \quad \rho_n = \frac{\lambda_1 \lambda_2 \cdots \lambda_{n-1}}{\mu_1 \mu_2 \cdots \mu_{n-1}}, \quad n = 2, 3, \ldots, N, \tag{4.3}$$

while π_n is simply related to ρ_n by

$$\pi_n = \frac{\mu_1}{\mu_n} \rho_n, \quad n = 1, 2, \ldots, N. \tag{4.4}$$

These relations hold for any birth–death process on the finite state space $\{0, 1, \ldots, N\}$ with an absorbing state at the origin. In particular, they can be used for the SIS model, which is defined by the relations $\lambda_n = \lambda(1 - n/N)n = \mu R_0(1 - n/N)n$, and $\mu_n = \mu n$, given in (2.14) and (2.15). We get

$$\rho_n = \frac{\lambda_1 \lambda_2 \cdots \lambda_{n-1}}{\mu_1 \mu_2 \cdots \mu_{n-1}} = \left(1 - \frac{1}{N}\right)\left(1 - \frac{2}{N}\right) \cdots \left(1 - \frac{n-1}{N}\right) R_0^{n-1}, \quad n = 2, 3, \ldots, N. \tag{4.5}$$

This leads to the explicit expression

$$\rho_n = \frac{1}{R_0} \frac{N!}{(N-n)!} \left(\frac{R_0}{N}\right)^n, \quad n = 1, 2, \ldots, N. \tag{4.6}$$

We note from this expression that the ratio ρ_{n+1}/ρ_n can be written as follows:

$$\frac{\rho_{n+1}}{\rho_n} = \left(1 - \frac{n}{N}\right) R_0, \quad n = 1, 2, \ldots, N - 1. \tag{4.7}$$

This shows that R_0 has a strong influence on the n-dependence of the quantities ρ_n. Thus, the ρ_n are decreasing in n for all n if $R_0 \leq 1$. (We assume always that $R_0 > 0$.) On the other hand, if $R_0 > 1$, then the quantities ρ_n increase in n up to an n-value close to $\bar{\mu}_1$, where

$$\bar{\mu}_1 = \frac{R_0 - 1}{R_0} N, \tag{4.8}$$

and decrease thereafter. We shall see later that $\bar{\mu}_1$ corresponds to the expected number of infected individuals in stationarity in both of the auxiliary processes, and also in quasi-stationarity in the original process, when $R_0 > 1$. To be specific, we find that ρ_n takes its maximum value for $n = [\bar{\mu}_1] + 1$ if $\bar{\mu}_1$ is not an integer, and that there are two equally large maximum values, taken at $n = \bar{\mu}_1$ and at $n = \bar{\mu}_1 + 1$, if $\bar{\mu}_1$ is an integer. Here, we use brackets around a real number x to denote the largest integer smaller than or equal to x. We note that $\bar{\mu}_1$ is the positive endemic infection level in the deterministic model. It corresponds to the carrying capacity in the population growth version of the model.

Furthermore, using (4.4) and $\mu_n = \mu n$, we get

$$\pi_n = \frac{\mu_1}{\mu_n} \rho_n = \frac{1}{n} \rho_n, \quad n = 1, 2, \ldots, N. \tag{4.9}$$

By using the expression for ρ_n in (4.6) we get

$$\pi_n = \frac{1}{nR_0} \frac{N!}{(N-n)!} \left(\frac{R_0}{N} \right)^n, \quad n = 1, 2, \ldots, N. \tag{4.10}$$

The n-dependence of π_n is similar to the n-dependence of ρ_n. The ratio of adjacent values equals

$$\frac{\pi_{n+1}}{\pi_n} = \frac{n}{n+1} \left(1 - \frac{n}{N} \right) R_0, \quad n = 1, 2, \ldots, N-1. \tag{4.11}$$

From this relation we find that π_n decreases in n for all n if $R_0 \leq 1$. If $R_0 > 2$, then π_n increases in n for n-values that are smaller than a value that can be written xN, where x is the larger of the two roots of the equation $x^2 - (R_0 - 1)x/R_0 + 1/(NR_0) = 0$, and decreases for larger n-values. For large N, this root is asymptotically equal to $\bar{\mu}_1 - 1/(R_0 - 1)$. If $1 < R_0 < 2$, then the n-dependence is similar, with the one exception that π_n decreases as a function of n for one or a few n-values larger than 1.

We use $I^{(0)}$ and $I^{(1)}$ to denote random variables with distributions $\mathbf{p}^{(0)}$ and $\mathbf{p}^{(1)}$, respectively. A simple relation between the two distributions $\mathbf{p}^{(0)}$ and $\mathbf{p}^{(1)}$ is derived. From (3.7) and (4.9) we get

$$p_n^{(0)} = \pi_n p_1^{(0)} = \frac{1}{n} \rho_n p_1^{(0)}, \quad n = 1, 2, \ldots, N, \tag{4.12}$$

and therefore

$$EI^{(0)} = \sum_{n=1}^{N} n p_n^{(0)} = \sum_{n=1}^{N} \rho_n p_1^{(0)} = \frac{p_1^{(0)}}{p_1^{(1)}}. \tag{4.13}$$

We conclude that the relation between the two stationary distributions can be expressed as follows:

$$\frac{p_n^{(0)}}{p_n^{(1)}} = \frac{\pi_n p_1^{(0)}}{\rho_n p_1^{(1)}} = \frac{1}{n} \frac{p_1^{(0)}}{p_1^{(1)}} = \frac{1}{n} EI^{(0)}, \quad n = 1, 2, \ldots, N. \tag{4.14}$$

This shows in particular that the probabilities $p_n^{(0)}$ are larger than $p_n^{(1)}$ whenever $n < EI^{(0)}$, and smaller when n exceeds $EI^{(0)}$.

We show also that the expectation of $I^{(0)}$ is smaller than the expectation of $I^{(1)}$. From (4.14) we get

$$\mathrm{E}I^{(0)}p_n^{(1)} = np_n^{(0)}, \quad n = 1, 2, \ldots, N, \tag{4.15}$$

which gives

$$\mathrm{E}I^{(0)}np_n^{(1)} = n^2 p_n^{(0)}, \quad n = 1, 2, \ldots, N. \tag{4.16}$$

We introduce $\mathrm{V}I^{(0)}$ to denote the variance of $I^{(0)}$. Summation over n from 1 to N gives

$$\mathrm{E}I^{(0)}\mathrm{E}I^{(1)} = \mathrm{V}I^{(0)} + \left(\mathrm{E}I^{(0)}\right)^2, \tag{4.17}$$

which leads to

$$\mathrm{V}I^{(0)} = \mathrm{E}I^{(0)}\left(\mathrm{E}I^{(1)} - \mathrm{E}I^{(0)}\right). \tag{4.18}$$

A consequence of this is that

$$\mathrm{E}I^{(0)} < \mathrm{E}I^{(1)}, \tag{4.19}$$

as claimed above, since clearly the variance of $I^{(0)}$ is positive.

Explicit expressions for the two stationary distributions $\mathbf{p}^{(1)}$ and $\mathbf{p}^{(0)}$ follow from (4.6) and (4.10). We get

$$p_n^{(1)} = \frac{1}{R_0}\frac{N!}{(N-n)!}\left(\frac{R_0}{N}\right)^n p_1^{(1)}, \quad n = 1, 2, \ldots, N, \tag{4.20}$$

and

$$p_n^{(0)} = \frac{1}{nR_0}\frac{N!}{(N-n)!}\left(\frac{R_0}{N}\right)^n p_1^{(0)}, \quad n = 1, 2, \ldots, N, \tag{4.21}$$

where

$$p_1^{(1)} = \frac{R_0/N!}{\sum_{n=1}^{N}(R_0/N)^n/(N-n)!}, \tag{4.22}$$

and

$$p_1^{(0)} = \frac{R_0/N!}{\sum_{n=1}^{N}(R_0/N)^n/(n(N-n)!)}. \tag{4.23}$$

The quasi-stationary distribution $\mathbf{q}$ and the two stationary distributions $\mathbf{p}^{(1)}$ and $\mathbf{p}^{(0)}$ are plotted in Fig. 4.1 for five different values of R_0, and with $N = 100$. The stationary distributions $\mathbf{p}^{(0)}$ and $\mathbf{p}^{(1)}$ are computed from the expressions in (4.20) and (4.21), while the quasi-stationary distribution $\mathbf{q}$ is computed by using the iteration procedure (3.32)–(3.36). The Maple procedures used for these computations are given in Appendix B.

The plots show that the three distributions behave very differently for different values of R_0. We shall identify three parameter regions where approximations of all three distributions show qualitatively different behaviors.

The plots also indicate that the approximations of $\mathbf{q}$ provided by $\mathbf{p}^{(1)}$ and $\mathbf{p}^{(0)}$ are not entirely satisfactory. For $R_0 = 1.5$ we observe that the body of $\mathbf{q}$ appears to be well approximated by $\mathbf{p}^{(0)}$ and reasonably well also by $\mathbf{p}^{(1)}$. The fact that $\mathbf{p}^{(0)}$ gives a better approximation of q than $\mathbf{p}^{(1)}$ for $R_0 > 1$ was noted already by Kryscio

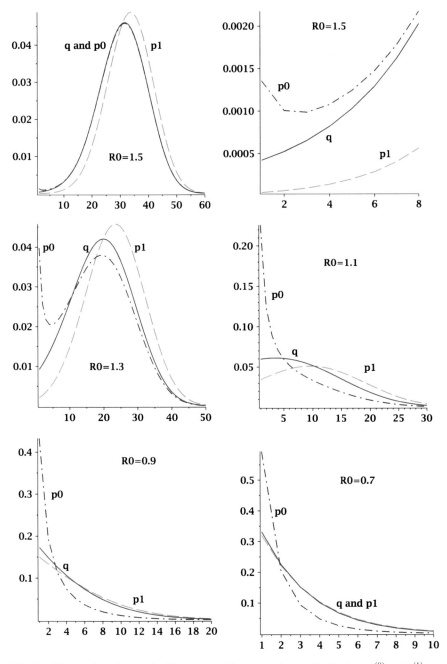

Fig. 4.1 The quasi-stationary distribution **q** and the two stationary distributions $\mathbf{p}^{(0)}$ and $\mathbf{p}^{(1)}$ are plotted for $N = 100$ and five different R_0-values. Note that all the distributions are discrete. Thus, the curve for e.g. $\mathbf{p}^{(0)}$ gives the probability $p_k^{(0)}$ as a function of k. The distributions are seen to depend strongly on R_0. Approximations of them are given in Chaps. 7, 9, and 11

and Lefèvre [39]. However, this difference is not of importance for our study. We shall see later that the difference between the bodies of $\mathbf{p}^{(1)}$ and $\mathbf{p}^{(0)}$ for $R_0 > 1$ is of small order asymptotically. Next, we observe that the important left tail of $\mathbf{q}$ is poorly approximated by both of the stationary distributions $\mathbf{p}^{(1)}$ and $\mathbf{p}^{(0)}$ when $R_0 = 1.5$. Improved approximations are required here in order to get a satisfactory approximation of the expected time to extinction from quasi-stationarity, since information about this expectation lies in the probability q_1.

The story is quite different when R_0 takes values close to one. We see for $R_0 = 1.3$ and $R_0 = 1.1$ that both $\mathbf{p}^{(1)}$ and $\mathbf{p}^{(0)}$ give poor approximations of $\mathbf{q}$, and that neither of them seems to be preferred over the other one. Here we are dealing with the so-called transition region in parameter space. It causes the most intricate problems in our search for approximations of the quasi-stationary distribution.

Finally, the plots in Fig. 4.1 show that $\mathbf{p}^{(1)}$ appears to give a much better approximation of the quasi-stationary distribution $\mathbf{q}$ than $\mathbf{p}^{(0)}$ in the two cases when $R_0 = 0.9$ and $R_0 = 0.7$. This will also be supported by our further studies.

We conclude from these observations that it is necessary to derive approximations of the quasi-stationary distribution $\mathbf{q}$ that provide improvements over the approximations given by the two stationary distributions $\mathbf{p}^{(1)}$ and $\mathbf{p}^{(0)}$. The remainder of the monograph is devoted to this task.

The first steps on this path give approximations of the stationary distribution $\mathbf{p}^{(1)}$ in Chaps. 6 and 7, and of the stationary distribution $\mathbf{p}^{(0)}$ in Chaps. 8 and 9. After that, we devote Chap. 10 to derivations of approximations of $\Psi(\mathbf{p}^{(1)})$ in all three parameter regions, of $\Psi(\mathbf{p}^{(0)})$ when $R_0 > 1$, and of $\Psi^n(\mathbf{p}^{(1)})$ in the transition region near $R_0 = 1$. These results lead to approximations of the quasi-stationary distribution $\mathbf{q}$ in each of the three parameter regions in Chap. 11. The final approximation step is taken in Chap. 13, where we derive an approximation of the quasi-stationary distribution $\mathbf{q}$ that is uniformly valid across all three of the parameter regions.

The approximations in Chaps. 6, 8, and 10 make use of approximations involving the normal distribution function Φ and the normal density function φ. Results of this type are derived in Chap. 5.

It is easy to verify that the transition rates λ_n and μ_n are proportional to the innocent parameter μ, but that ρ_n and π_n are independent of μ. This has the consequence that μ has no influence on the stationary distributions $\mathbf{p}^{(0)}$ and $\mathbf{p}^{(1)}$, nor on the quasi-stationary distribution $\mathbf{q}$. Thus, all three of these distributions are determined by only two parameters, namely the essential parameters N and R_0.

Chapter 5
Some Approximations Involving the Normal Distribution

Approximations play a central role in this monograph. The subject area of asymptotic analysis is therefore important. It contains a number of powerful results and ideas. A classic reference to this field is the book by Olver [54].

We give a brief introduction to concepts and notation in asymptotic analysis. Let $f(x)$ be a function that we wish to study, and $g(x)$ a known function. Assume that $f(x)/g(x) \to 1$ as $x \to \infty$. We say then that g is *an asymptotic approximation* to f, and write

$$f(x) \sim g(x), \quad x \to \infty, \tag{5.1}$$

If $f(x)/g(x) \to 0$ as $x \to \infty$, then we say that f is *of order less than* g, and write

$$f(x) = o(g(x)), \quad x \to \infty. \tag{5.2}$$

If $|f(x)/g(x)|$ is bounded, then we say that f is *of order not exceeding* g, and write

$$f(x) = O(g(x)), \quad x \to \infty. \tag{5.3}$$

Special cases of these definitions are: $f(x) = o(1)$ as $x \to \infty$, meaning that $f(x) \to 0$ as $x \to \infty$, and $f(x) = O(1)$ as $x \to \infty$, meaning that $|f(x)|$ is bounded as $x \to \infty$.

Assume that $f(x) = \sum_{k=0}^{\infty} b_k/x^k$ is a formal power series, which may be convergent or divergent. Let $R_n(x)$ be the difference between $f(x)$ and the n'th partial sum of the series. Thus we have

$$f(x) = b_0 + \frac{b_1}{x} + \frac{b_2}{x^2} + \cdots + \frac{b_{n-1}}{x^{n-1}} + R_n(x). \tag{5.4}$$

Let us assume that for each fixed n

$$R_n(x) = O\left(\frac{1}{x^n}\right), \quad x \to \infty. \tag{5.5}$$

I. Nåsell, *Extinction and Quasi-stationarity in the Stochastic Logistic SIS Model*,
Lecture Notes in Mathematics 2022, DOI 10.1007/978-3-642-20530-9_5,
© Springer-Verlag Berlin Heidelberg 2011

Then we say that the series $\sum_{k=1}^{\infty} b_k/x^k$ is *an asymptotic expansion* of $f(x)$, and write

$$f(x) \sim b_0 + \frac{b_1}{x} + \frac{b_2}{x^2} + \dots, \quad x \to \infty. \tag{5.6}$$

Clearly, the first term in an asymptotic expansion of $f(x)$ provides an asymptotic approximation of $f(x)$. Furthermore, additional terms of an asymptotic expansion provide improved approximations. We shall refer to them also as asymptotic approximations. We note that the symbol $\sim$ has different meanings in (5.1) and (5.6).

We shall in several instances encounter distributions with mean μ and standard deviation σ. It is useful to describe the possible state values as belonging either to the body or the near-end tails or the far-end tails of the distribution. We do this by introducing two functions $\hat{y}$ and $\hat{z}$ as follows:

$$\hat{y}(x) = \frac{x - \mu}{\sigma}, \tag{5.7}$$

$$\hat{z}(x) = \frac{x - \mu}{\sigma^2} = \frac{\hat{y}(x)}{\sigma}. \tag{5.8}$$

We shall throughout be concerned with situations where the standard deviation σ is large. We shall therefore strive for approximations that are asymptotic as $\sigma \to \infty$. We regard a point x as belonging to the body of the distribution if the difference between x and the mean μ is a finite number of standard deviations σ. Translating this into asymptotic language, we find that x belongs to the body if $\hat{y}(x) = O(1)$ as $\sigma \to \infty$. Any point that does not lie in the body of the distribution belongs to one of the tails. We give a description of a tail point when all real values of x are possible. Any tail point x is then characterized by the criterion that $|\hat{y}(x)| \to \infty$ as $\sigma \to \infty$. We use the function $\hat{z}$ to distinguish between near-end tails and far-end tails, again using asymptotic concepts. Thus, x belongs to a near-end tail if $|\hat{y}(x)| \to \infty$ while $\hat{z}(x) = o(1)$ as $\sigma \to \infty$, and it belongs to a far-end tail if $|\hat{y}(x)| \to \infty$ while $\hat{z}(x)$ is of the order of one or larger as $\sigma \to \infty$. An alternative way of describing the body or near-end tail is through the condition $\hat{y}(x) = o(\sigma)$. This means that $\hat{y}(x)$ is allowed to grow with σ, but at a lower rate. The distinction between the three cases of body, near-end tail, and far-end tail will be of importance at several points in the monograph. Examples of situations where x lies in the body, or a near-end tail, or a far-end tail are given by $x = \mu + k\sigma$, $x = \mu + k\sigma^{3/2}$, and $x = \mu + k\sigma^2$, respectively, where $k \neq 0$ is bounded.

The definition of a tail point is different in the case when x is restricted to positive values, as is usually the case in this monograph. The left tail is then characterized by all sufficiently small values of x, independent of the value of μ. The restriction to small x-values does not imply that x must be bounded, only that its growth rate is restricted. We shall see several examples in the following chapters.

Our concern throughout this monograph is with discrete distributions, where there are finitely many possible values that are integral and positive, as e.g.

$k = 1, 2, \ldots, N$. The distribution is then given by a probability vector $\mathbf{p} = (p_1, p_2, \ldots, p_N)$, where $p_k = P\{X = k\}$. Such distributions can in many cases be approximated by the aid of continuous normal random variables with known mean μ and known standard deviation σ. This means that the probability p_k is approximated by an expression that contains the function $\varphi(\hat{y}(k))$, where φ is used to denote the normal density function $\varphi(y) = \exp(-y^2/2)/\sqrt{2\pi}$. We shall also use the normal distribution function $\Phi(y) = \int_{-\infty}^{y} \varphi(u)\,du$. Furthermore, the reciprocal φ^* of the normal density function φ will play a role. It is defined by

$$\varphi^*(y) = \frac{1}{\varphi(y)} = \sqrt{2\pi}\exp(y^2/2), \tag{5.9}$$

while its integral Φ^* is equal to

$$\Phi^*(y) = \int_0^y \varphi^*(x)\,dx. \tag{5.10}$$

The remainder of the chapter is disposed as follows: Sect. 5.1 is used to study two functions of x formed by the normal density function φ and its reciprocal φ^*, both evaluated at the argument $\hat{y}(k + x)$. The powerful results here are asymptotic approximations that are simple in form and nonlinear and definite improvements over the linear approximations provided by the first two terms of Maclaurin expansions. In Sect. 5.2 we give similar results for the normal distribution function Φ and for the integral Φ^* of the reciprocal of the normal density function. The ratio between the normal distribution function Φ and the normal density function φ is of considerable importance in the later part of our study. Well-known asymptotic properties are given in Sect. 5.3, together with asymptotic properties of the ratio between the two functions Φ^* and φ^*. The most important results of the chapter are given in Sects. 5.4 and 5.5. The first of these sections deals with sums of normal densities, and the second one with sums of reciprocal normal densities. The results show a rather intricate and unexpected behavior, with different behaviors in body, near-end tails, and far-end tails. The so-called Cox continuity correction has been introduced to deal with the sums of normal densities that we study in Sect. 5.4. We show in Sect. 5.6 that our approximation can be used to derive an approximation of the Cox continuity correction. The approximation takes values in the unit interval $(0,1)$. It extends the well-known approximation of $1/2$ from the body and the near-end tails to the far-end tails of the distribution.

As already indicated, we deal with various approximations involving the normal distribution in this chapter. In each such case, we present both quite general asymptotic expansions, and much simpler asymptotic approximations containing only one or two terms. The asymptotic expansions that we give in Sects. 5.1 and 5.2 are derived, while those in Sects. 5.4 and 5.5 are left as open problems. It turns out that the applications in later chapters require only the simpler asymptotic approximations. The reason for considering the more demanding asymptotic expansions is that this approach gives additional insight into the problems studied, and also that additional terms may be required if one wants to derive improved approximations of the various quantities that are of interest in the study of the SIS model, or of other stochastic models.

5.1 Approximations of the Normal Density Function φ and of Its Reciprocal φ^* at the Argument $\hat{y}(k+x)$

In this section we study $\varphi(\hat{y}(k+x))$ and $\varphi^*(\hat{y}(k+x))$ as functions of x for fixed values of k, μ, and σ. We derive asymptotic expansions of both that are valid for large σ when x is of smaller order than σ. These results are contained in the theorem below. In what follows we shall use the asymptotic approximations that are formed by including one term only from these asymptotic expansions, and associated bounds. These consequences of the theorem are given in the corollary that follows.

Theorem 5.1 *The functions $\varphi(\hat{y}(k+x))$ and $\varphi^*(\hat{y}(k+x))$ obey the following asymptotic expansions:*

$$\varphi(\hat{y}(k+x)) \sim \varphi(\hat{y}(k)) \exp(-x\hat{z}(k)) \sum_{m=0}^{\infty} \frac{1}{m!} \left(-\frac{x^2}{2\sigma^2} \right)^m, \quad x = o(\sigma), \quad \sigma \to \infty,$$

$$\tag{5.11}$$

$$\varphi^*(\hat{y}(k+x)) \sim \varphi^*(\hat{y}(k)) \exp(x\hat{z}(k)) \sum_{m=0}^{\infty} \frac{1}{m!} \left(\frac{x^2}{2\sigma^2} \right)^m, \quad x = o(\sigma), \quad \sigma \to \infty.$$

$$\tag{5.12}$$

Proof. The derivations of these asymptotic expansions are easy. All that is required is to note that the ratio $\varphi(\hat{y}(k+x))/\varphi(\hat{y}(k))$ can be expressed as follows:

$$\frac{\varphi(\hat{y}(k+x))}{\varphi(\hat{y}(k))} = \exp\left(\frac{(\hat{y}(k))^2}{2} - \frac{(\hat{y}(k+x))^2}{2} \right)$$

$$= \exp\left(-\frac{x(k-\mu)}{\sigma^2} - \frac{x^2}{2\sigma^2} \right) = \exp(-x\hat{z}(k)) \exp\left(-\frac{x^2}{2\sigma^2} \right). \tag{5.13}$$

By replacing the second factor in the last of these expressions by its Maclaurin expansion, we are led to the explicit expression

$$\varphi(\hat{y}(k+x)) = \varphi(\hat{y}(k)) \exp(-x\hat{z}(k)) \sum_{m=0}^{\infty} \frac{1}{m!} \left(-\frac{x^2}{2\sigma^2} \right)^m. \tag{5.14}$$

This is an asymptotic expansion if x is of smaller order that σ, as is required in (5.11).

The result in (5.12) follows since $\varphi^*(y) = 1/\varphi(y)$. $\square$

Corollary 5.1 *The functions* $\varphi(\hat{y}(k+x))$ *and* $\varphi^*(\hat{y}(k+x))$ *obey the following asymptotic approximations and bounds:*

$$\varphi(\hat{y}(k+x)) \sim \varphi(\hat{y}(k)) \exp(-x\hat{z}(k)), \quad x = o(\sigma), \quad \sigma \to \infty, \tag{5.15}$$

$$\varphi^*(\hat{y}(k+x)) \sim \varphi^*(\hat{y}(k)) \exp(x\hat{z}(k)), \quad x = o(\sigma), \quad \sigma \to \infty, \tag{5.16}$$

$$\varphi(\hat{y}(k+x)) < \varphi(\hat{y}(k)) \exp(-x\hat{z}(k)), \tag{5.17}$$

$$\varphi^*(\hat{y}(k+x)) > \varphi^*(\hat{y}(k)) \exp(x\hat{z}(k)). \tag{5.18}$$

Proof. The asymptotic approximations follow from the asymptotic expansions in the theorem above by including one term only.

The bounds follow since $\exp(-x^2/(2\sigma^2)) < 1$ and $\exp(x^2/(2\sigma^2)) > 1$ without restrictions on x or σ. $\square$

The corollary shows that the functions $\varphi(\hat{y}(k+x))$ and $\varphi^*(\hat{y}(k+x))$ are approximately linear in x when $|x| \leq 1$ and k lies in the body or the near-end tails of the distribution with mean μ and standard deviation σ, since then $\hat{z}(k) = o(1)$, which implies that $\exp(-x\hat{z}(k)) \sim 1 - x\hat{z}(k)$. However, the corollary also shows the important result that this linearity no longer holds when k lies in the far-end tails of the same distribution.

In similarity to this, we shall find in Sects. 5.4 and 5.5 that approximations of sums of normal density functions and of reciprocals of normal density functions take different forms in near-end and far-end tails. The function $\hat{z}(k)$ carries information about this in a compact way.

We note that the asymptotic approximations in the above corollary are more general than the linear approximations that one gets by including the first two terms in the Maclaurin expansions of the same function, since they are valid for a larger interval of x-values, and since they are not limited to linear functions.

5.2   Approximations of the Normal Distribution Function Φ and of the Integral Φ^* of the Reciprocal of the Normal Density Function at the Argument $\hat{y}(k+x)$

This section deals with the normal distribution function $\Phi(\hat{y}(k+x))$ and with the integral $\Phi^*(\hat{y}(k+x))$ of the reciprocal of the normal density, both as functions of x, for fixed values of k, μ, and σ. To describe our results, we define the function G_2 as follows:

$$G_2(x,z) = \begin{cases} \dfrac{1 - \exp(-xz)}{z}, & z \neq 0, \\ x, & z = 0. \end{cases} \tag{5.19}$$

The function G_2 is defined and analytic for all real values of x and z. Clearly, it is identically equal to 0 if $x = 0$. If $x \neq 0$, then $G_2(x, z)$ has the same sign as x, and is a monotonically decreasing function of z. We note also that it can be expressed in terms of an incomplete gamma function as follows:

$$G_2(x, z) = \frac{\gamma(1, xz)}{z}, \tag{5.20}$$

where γ denotes the lower incomplete gamma function defined by

$$\gamma(a, y) = \int_0^y t^{a-1} \exp(-t) \, dt. \tag{5.21}$$

Here, we use the notation in Abramowitz and Stegun [1], Formula 6.5.2, or, equivalently, Sect. 8.2.1 in Olver et al. [55]. The partial derivatives of G_2 with respect to z can be expressed in terms of the lower incomplete gamma function as follows:

$$\frac{\partial^m}{\partial z^m} G_2(x, z) = (-1)^m \frac{\gamma(m+1, xz)}{z^{m+1}}, \quad m = 0, 1, 2, \ldots. \tag{5.22}$$

As in the previous section, we give both asymptotic expansions and much simpler asymptotic approximations of both functions $\Phi(\hat{y}(k+x))$ and $\Phi^*(\hat{y}(k+x))$. The asymptotic expansions are given in the following theorem:

Theorem 5.2 *The functions* $\Phi(\hat{y}(k+x))$ *and* $\Phi^*(\hat{y}(k+x))$ *obey the following asymptotic expansions:*

$$\Phi(\hat{y}(k+x)) \sim \Phi(\hat{y}(k)) + \frac{1}{\sigma} \varphi(\hat{y}(k)) \sum_{m=0}^{\infty} \frac{1}{m!} \left(-\frac{1}{2\sigma^2} \right)^m \frac{\partial^{2m}}{\partial z^{2m}} G_2(x, \hat{z}(k)),$$

$$x = o(\sigma), \quad \sigma \to \infty, \quad (5.23)$$

$$\Phi^*(\hat{y}(k+x)) \sim \Phi^*(\hat{y}(k)) + \frac{1}{\sigma} \varphi^*(\hat{y}(k)) \sum_{m=0}^{\infty} \frac{1}{m!} \left(\frac{1}{2\sigma^2} \right)^m \frac{\partial^{2m}}{\partial z^{2m}} G_2(x, \hat{z}(k)),$$

$$x = o(\sigma) \quad \sigma \to \infty, \quad (5.24)$$

Proof. To derive (5.23), we integrate the asymptotic expansion in (5.11) with respect to x. Since the sum in (5.11) converges uniformly, the integral of the sum equals the sum of the integrals:

$$\sigma\Phi(\hat{y}(k+x)) - \sigma\Phi(\hat{y}(k)) \sim \varphi(\hat{y}(k)) \sum_{m=0}^{\infty} \frac{1}{m!} \left(-\frac{1}{2\sigma^2}\right)^m \int_0^x \exp(-t\hat{z}(k))t^{2m}\,dt,$$

$$x = o(\sigma), \quad \sigma \to \infty. \quad (5.25)$$

We evaluate the integral:

$$\int_0^x t^{2m}\exp(-tz)\,dt = \frac{1}{z^{2m+1}}\int_0^{xz} y^{2m}\exp(-y)\,dy = \frac{1}{z^{2m+1}}\gamma(2m+1,xz)$$

$$= \frac{\partial^{2m}}{\partial z^{2m}}G_2(x,z), \quad (5.26)$$

using (5.22). Insertion of this expression for the integral establishes (5.23). The derivation of (5.24) is similar.   □

We quote the asymptotic approximations of $\Phi(\hat{y}(k+x))$ and $\Phi^*(\hat{y}(k+x))$ that follow from the asymptotic expansions given in the above theorem by including two terms only. The results are:

Corollary 5.2

$$\Phi(\hat{y}(k+x)) \sim \Phi(\hat{y}(k)) + \frac{1}{\sigma}\varphi(\hat{y}(k))G_2(x,\hat{z}(k)), \quad x = o(\sigma), \quad \sigma \to \infty, \quad (5.27)$$

$$\Phi^*(\hat{y}(k+x)) \sim \Phi^*(\hat{y}(k)) + \frac{1}{\sigma}\varphi^*(\hat{y}(k))G_2(x,\hat{z}(k)), \quad x = o(\sigma), \quad \sigma \to \infty. \quad (5.28)$$

It is easy to see from the definition of G_2 that these approximations are asymptotically linear in x if $x\hat{z}(k) = o(1)$. If we consider $|x| \leq 1$, this translates to the condition $\hat{z}(k) = o(1)$ on k. We note also that the approximations for $|x| \leq 1$ are non-linear in the far-end tails where $\hat{z}(k)$ is of the order of one or larger.

5.3   Approximations of the Ratios Φ/φ and Φ^*/φ^*

Two functions related to the normal distribution will play important roles in what follows. We denote them by H_1 and H_1^*, respectively. The first of these functions is defined as the ratio of the normal distribution function Φ to the normal density function φ:

$$H_1(y) = \frac{\Phi(y)}{\varphi(y)}, \quad (5.29)$$

while the second of the two functions is defined as the ratio between the integral of the reciprocal of the normal density function Φ^* and the reciprocal of the normal density φ^*:

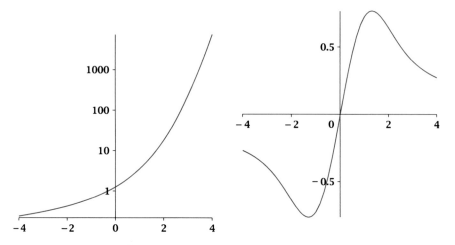

Fig. 5.1 The function $H_1(y)$ is shown as a function of y in the *left* figure, and the function $H_1^*(y)$ is shown as a function of y in the *right* figure

$$H_1^*(y) = \frac{\Phi^*(y)}{\varphi^*(y)}. \tag{5.30}$$

The two functions are illustrated in Fig. 5.1. We give some properties of these functions in this section, as a preparation for their use in later parts of the monograph.

The functions H_1 and H_1^* are closely related to Mills' ratio R for the normal distribution and to Dawson's integral F, respectively. Mills' ratio for the normal distribution is defined by

$$R(y) = \frac{1 - \Phi(y)}{\varphi(y)}. \tag{5.31}$$

It follows therefore that

$$H_1(y) = R(-y). \tag{5.32}$$

Similarly, Dawson's integral F is defined by

$$F(y) = \exp(-y^2) \int_0^y \exp(t^2)\, dt. \tag{5.33}$$

This leads to the following relation between H_1^* and F:

$$H_1^*(y) = \sqrt{2}\, F\left(\frac{y}{\sqrt{2}}\right). \tag{5.34}$$

To describe the results in this section, we define a sequence of positive coefficients a_k by

$$a_k = \frac{(2k)!}{k!2^k}, \quad k = 0, 1, 2, \dots. \tag{5.35}$$

The first few coefficients are $a_0 = 1$, $a_1 = 1$, $a_2 = 1 \cdot 3 = 3$, $a_3 = 1 \cdot 3 \cdot 5 = 15$, $a_4 = 1 \cdot 3 \cdot 5 \cdot 7 = 105$. Note also that $a_k = (2k-1)a_{k-1}$ for $k \geq 1$. The main results in this section are asymptotic expansions of H_1 and H_1^*. They are given in the following theorem:

Theorem 5.3 *The functions $H_1(y) = \Phi(x)/\varphi(x)$ and $H_1^*(x) = \Phi^*(x)/\varphi^*(x)$ have the following asymptotic expansions:*

$$H_1(y) \sim \sum_{k=0}^{\infty} (-1)^{k+1} \frac{a_k}{y^{2k+1}}, \quad y \to -\infty, \tag{5.36}$$

$$H_1(y) \sim \frac{1}{\varphi(y)} + \sum_{k=0}^{\infty} (-1)^{k+1} \frac{a_k}{y^{2k+1}}, \quad y \to +\infty, \tag{5.37}$$

$$H_1^*(y) \sim \sum_{k=0}^{\infty} \frac{a_k}{y^{2k+1}}, \quad |y| \to \infty, \tag{5.38}$$

Proof. We use the asymptotic expansion given in Formula 7.1.23 in Abramowitz and Stegun [1], or in Sect. 7.12.1 in Olver et al. [55] for the complementary error function erfc with complex argument z. The asymptotic expansions given in the theorem follow after noting that $\Phi(x) = 1 - \text{erfc}(x/\sqrt{2})/2$. (We note that the function Φ defined in the Digital Library of Mathematical Functions with release date 2010-05-07 differs from the function Φ that we use here.)   □

The following asymptotic approximations follow immediately from the above theorem:

Corollary 5.3 *The two functions H_1 and H_1^* have the following asymptotic approximations:*

$$H_1(y) \sim \frac{1}{\varphi(y)}, \quad y \to +\infty, \tag{5.39}$$

$$H_1(y) \sim -\frac{1}{y}, \quad y \to -\infty, \tag{5.40}$$

$$H_1^*(y) \sim \frac{1}{y}, \quad y \to \pm\infty. \tag{5.41}$$

The asymptotic expansion for $H_1(y)$ given here for $y \to -\infty$ is classic. Some version of it was known already to Laplace, although the concept of asymptotic expansion was formulated much later by Poincaré.

The function H_1 is a monotonically increasing function of y. It makes a continuous transition from small positive values when the argument y is negative with a large magnitude to large positive values when the argument y is positive and large, as seen in Fig. 5.1, and also shown by the asymptotic approximations of $H_1(y)$ as $y \to \pm\infty$ given in the corollary. Thus, it behaves qualitatively very differently in these two cases. We shall see as we proceed that these two behaviors of $H_1(y)$ correspond to two qualitatively different behaviors of the stationary distribution $\mathbf{p}^{(1)}$ for the SIS model in the two parameter regions where R_0 is distinctly below one and above one, respectively. We shall also find that the function H_1 can be used to represent a continuous transition between these very different distributional forms for $\mathbf{p}^{(1)}$. Similar behaviors will also be found for the stationary distribution $\mathbf{p}^{(0)}$ and for the quasi-stationary distribution $\mathbf{q}$ for the SIS model. Continuous transitions between qualitatively different distributional forms for these two distributions will also be established. The functions that describe them will be defined as we proceed. They show similarities to the function H_1. They will be denoted H_0 and H, respectively.

Furthermore, successive asymptotic approximations of $H_1(y)$ as $y \to -\infty$ provide upper and lower bounds for $H_1(y)$ when $y < 0$. To describe them, we define a sequence of functions $A_m(y)$ as follows:

$$A_m(y) = \sum_{k=0}^{m} (-1)^{k+1} \frac{a_k}{y^{2k+1}} = -\frac{1}{y} + \frac{1}{y^3} - \frac{1\cdot 3}{y^5} + \frac{1\cdot 3\cdot 5}{y^7} + \cdots + \frac{(-1)^{m+1}a_m}{y^{2m+1}},$$

$$m = 0, 1, 2, \ldots, \quad y < 0. \quad (5.42)$$

We note that $A_m(y)$ does not converge as $m \to \infty$. The function $A_m(y)$ is an upper bound for $H_1(y)$ when $y < 0$ if m is even, and a lower bound for $H_1(y)$ when $y < 0$ if m is odd. Thus, we have

$$H_1(y) < A_m(y), \quad m = 0, 2, 4, \ldots, \quad y < 0, \quad (5.43)$$

$$H_1(y) > A_m(y), \quad m = 1, 3, 5, \ldots, \quad y < 0. \quad (5.44)$$

A particular and useful upper bound of H_1 is given by

$$H_1(y) < A_0(y) = -\frac{1}{y}, \quad y < 0. \quad (5.45)$$

The two asymptotic approximations for $H_1(y)$ as $y \to \pm\infty$ can be combined into one formula. We introduce Θ to denote Heaviside's step function, which equals zero when its argument is negative, and one when its argument is positive. Thus,

$$\Theta(y) = \begin{cases} 1, & y > 0, \\ 0, & y < 0. \end{cases} \quad (5.46)$$

The two results can then be combined into the following formula:

$$H_1(y) \sim \frac{\Theta(y)}{\varphi(y)} + A_m(y), \quad m = 0, 1, 2, \ldots, \quad |y| \to \infty. \tag{5.47}$$

5.4   An Approximation of a Sum of Normal Densities

This section gives an asymptotic approximation of a sum of normal densities of the form $\frac{1}{\sigma} \sum_{j=-\infty}^{K} \varphi(\hat{y}(j))$. As before, $\hat{y}(j) = (j - \mu)/\sigma$, where μ and σ are known, and where σ is large.

Sums of normal densities will appear at a number of places in Chaps. 6, 8, and 10. Approximations of such sums are of high importance for the intricate results derived in these chapters. We prepare for these applications by giving a theorem below that contains an asymptotic approximation of such a sum. It turns out that this approximation takes different forms in the three different cases when the quantity K lies in the body, or a near-end tail, or a far-end tail, of a distribution with mean μ and standard deviation σ. We spell out these three cases in a corollary to the theorem. We cite also a conjectured asymptotic expansion, but we leave it as an open problem to prove that the conjecture is true.

The approximation uses a function G_1 defined as follows:

$$G_1(z) = \begin{cases} 1 - \dfrac{1}{z} + \dfrac{1}{\exp(z) - 1}, & z \neq 0, \\[2mm] \dfrac{1}{2}, & z = 0, \end{cases} \tag{5.48}$$

This function takes values in the open unit interval (0,1), is monotonically increasing, and approaches the values zero and one as the argument z approaches $-\infty$ and $+\infty$, respectively. Thus, it can be interpreted as a distribution function of a continuous random variable. Also, it is symmetric in the sense that $G_1(z) + G_1(-z) = 1$. It is straightforward to verify these properties of the function G_1. A plot is given in Fig. 5.2 at the end of this chapter.

The main result in this section is as follows:

Theorem 5.4 *The sum of normal densities $\frac{1}{\sigma} \sum_{j=-\infty}^{K} \varphi(\hat{y}(j))$ has the following asymptotic approximation:*

$$\frac{1}{\sigma} \sum_{j=-\infty}^{K} \varphi(\hat{y}(j)) \sim \Phi(\hat{y}(K)) + \frac{1}{\sigma} \varphi(\hat{y}(K)) [1 - G_1(\hat{z}(K))], \quad \sigma \to \infty. \tag{5.49}$$

This result is formally equivalent to an approximation that was contained in Nåsell [44], but with the important difference that we prove here that the approximation is asymptotic, while a scrutiny of the 1996 derivation shows that asymptoticity was not proved at that time. Our derivation of this important approximation is given below. As far as we know, it has not appeared before.

We add the conjecture that the same sum of normal densities actually obeys the following asymptotic expansion:

$$\frac{1}{\sigma} \sum_{j=-\infty}^{K} \varphi(\hat{y}(j)) \sim \Phi(\hat{y}(K)) + \frac{1}{\sigma} \varphi(\hat{y}(K)) \left[1 - \sum_{m=0}^{\infty} \frac{1}{m!} \left(-\frac{1}{2\sigma^2} \right)^m G_1^{(2m)}(\hat{z}(K)) \right],$$

$$\sigma \to \infty. \quad (5.50)$$

We leave it as an open problem to derive this result.

We use the result in (5.49) to derive three asymptotic approximations that are valid in the three cases when K lies in the body, or a near-end tail, or a far-end tail of the distribution with mean μ and standard deviation σ, respectively. The three approximations are given in the following corollary:

Corollary 5.4 *The sum of normal densities $\frac{1}{\sigma}\sum_{j=-\infty}^{K}\varphi(\hat{y}(j))$ has the following specific asymptotic approximations:*

$$\frac{1}{\sigma} \sum_{j=-\infty}^{K} \varphi(\hat{y}(j)) \sim \Phi(\hat{y}(K)) + \frac{1}{2\sigma} \varphi(\hat{y}(K)), \quad \hat{y}(K) = O(1), \quad \sigma \to \infty, \quad (5.51)$$

$$\frac{1}{\sigma} \sum_{j=-\infty}^{K} \varphi(\hat{y}(j)) \sim \Theta(\hat{y}(K)) - \frac{\varphi(\hat{y}(K))}{\hat{y}(K)}, \quad |\hat{y}(K)| \to \infty, \quad \hat{z}(K) = o(1), \quad \sigma \to \infty,$$

$$(5.52)$$

$$\frac{1}{\sigma} \sum_{j=-\infty}^{K} \varphi(\hat{y}(j)) \sim \Theta(\hat{y}(K)) - \frac{1}{\sigma} \frac{\varphi(\hat{y}(K))}{\exp(\hat{z}(K)) - 1}, \quad |\hat{y}(K)| \to \infty, \quad \sigma \to \infty. \quad (5.53)$$

In the latter two of these expressions we use Θ to denote Heaviside's step function defined in (5.46).

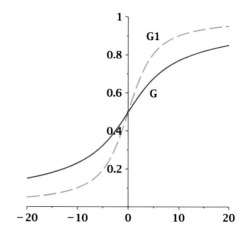

Fig. 5.2 The functions $G(z)$ and $G_1(z)$ are shown as functions of z. Both of them can be interpreted as distribution functions of real random variables with heavy tails and no moments

Clearly, the first one of these three approximations holds when K lies in the body of the distribution with mean μ and standard deviation σ, while the second one holds when K lies in a near-end tail of this distribution, and the third one holds when K lies in any tail, either near-end or far-end, of the same distribution.

We show first that the approximations in the corollary follow from the approximation given in the theorem.

The approximation (5.51) follows readily from the one in (5.49), since the condition for K to lie in the body of the distribution, $\hat{y}(K) = O(1)$, implies that $\hat{z}(K) = o(1)$, and therefore $G_1(\hat{z}(K)) \sim 1/2$.

The approximations in (5.52) and (5.53) both require K to lie in a tail of the distribution. The first term on the right-hand side of the approximation (5.49) is then either very close to zero (when K lies in a left tail) or very close to one (when K lies in a right tail). We can then use the following asymptotic approximation of $\Phi(y)$:

$$\Phi(y) \sim \Theta(y) - \frac{\varphi(y)}{y}, \quad |y| \to \infty. \tag{5.54}$$

It follows from the asymptotic approximation of $H_1(y)$ in (5.47) with $m = 0$ after multiplication by $\varphi(y)$.

By applying the approximation (5.54) to the first term on the right-hand side of (5.49), and also using the definition (5.48) of the function G_1, we find that (5.53) holds. The further simplification contained in (5.52) follows when K lies in a near-end tail. The condition $\hat{z}(K) = o(1)$ then implies that $\exp(\hat{z}(K)) \sim 1 + \hat{z}(K)$. Applying this in (5.53) leads to (5.52).

We turn now toward derivation of the asymptotic approximation of the sum of normal densities given in (5.49). We consider three cases that are treated separately. In the first case K lies in the body or a near-end tail so that $\hat{z}(K) = o(1)$. In the second case we take K in the right far-end tail, so that $\hat{z}(K) > 0$ and $\hat{z}(K)$ is of the order of one or larger. The third case, finally, deals with the left far-end tail, where $\hat{z}(K) < 0$ and $\hat{z}(K)$ is of order one or larger.

The approximation (5.49) to be proved simplifies somewhat in each of these three cases. In case $\hat{z}(K) = o(1)$, we find that $G_1(\hat{z}(K)) \sim 1/2$. It suffices therefore to prove

$$\frac{1}{\sigma} \sum_{j=-\infty}^{K} \varphi(\hat{y}(j)) \sim \Phi(\hat{y}(K)) + \frac{1}{2\sigma}\varphi(\hat{y}(K)), \quad \hat{z}(K) = o(1), \quad \sigma \to \infty, \tag{5.55}$$

in the first case. In the second case we use the approximation (5.54) of $\Phi(y)$ for $y \to \infty$ and the definition (5.48) of the function G_1 to deduce that (5.49) holds if we prove that

$$\frac{1}{\sigma} \sum_{j=-\infty}^{K} \varphi(\hat{y}(j)) \sim 1 - \frac{1}{\sigma} \varphi(\hat{y}(K)) \frac{1}{\exp(\hat{z}(K)) - 1},$$

$$\hat{z}(K) > 0, \quad \hat{z}(K) \text{ of order of 1 or larger,} \quad \sigma \to \infty. \quad (5.56)$$

Finally, in the third case we use (5.54) with $y \to -\infty$ and the definition of G_1 to show that it suffices to prove that

$$\frac{1}{\sigma} \sum_{j=-\infty}^{K} \varphi(\hat{y}(j)) \sim \frac{1}{\sigma} \varphi(\hat{y}(K)) \frac{1}{1 - \exp(\hat{z}(K))},$$

$$\hat{z}(K) < 0, \quad \hat{z}(K) \text{ of order of 1 or larger,} \quad \sigma \to \infty. \quad (5.57)$$

We proceed to show that (5.55) holds. We identify two different cases where different methods are used. First we consider a restriction to K-values for which $\hat{y}(K) < -1$, and make use of the Euler-Maclaurin formula. In the second case we take $\hat{y}(K) \geq -1$. The derivation of (5.55) then uses the result just derived for $\hat{y}(K) < -1$, and the asymptotic approximation of $\Phi(\hat{y}(K+x))$ given in in (5.27).

The Euler-Maclaurin formula provides a powerful way of deriving asymptotic approximations of sums of functions. Following Olver [54], it can be written as follows

$$\sum_{j=k}^{K} f(j) = \int_{k}^{K} f(x) \, dx + \frac{1}{2} [f(k) + f(K)] + S_f(K,M) - S_f(k,M) + \bar{R}_M,$$

$$M = 1, 2, \ldots, \quad k < K, \quad (5.58)$$

where

$$S_f(i,M) = \sum_{j=1}^{M-1} \frac{B_{2j}}{(2j)!} f^{(2j-1)}(i), \quad M = 1, 2, \ldots, \quad (5.59)$$

and the remainder term $\bar{R}_M$ is given by

$$\bar{R}_M = \frac{1}{(2M)!} \int_{k}^{K} (B_{2M} - B_{2M}(x - [x])) f^{(2M)}(x) \, dx, \quad M = 1, 2, \ldots \quad (5.60)$$

Since the integral for $\bar{R}_M$ contains even derivatives of f, this is an even-order derivative version of the Euler-Maclaurin formula. (There is also an odd-order derivative version, given by Apostol [5].) In the expressions above, B_m are Bernoulli numbers, and $B_m(x)$ are Bernoulli polynomials. The functions $B_m(x - [x])$ are periodic extensions of the Bernoulli polynomials $B_m(x)$, with period one. They are called Bernoulli periodic functions. The Bernoulli polynomials are defined as the coefficients of $z^m/m!$ in the Maclaurin series expansion of the function $z \exp(xz)/(\exp(z) - 1)$. Explicitly we have

$$\frac{z\exp(xz)}{\exp(z) - 1} = \sum_{m=0}^{\infty} B_m(x)\frac{z^m}{m!}, \quad |z| < 2\pi. \tag{5.61}$$

The first few Bernoulli polynomials are $B_0(x) = 1$, $B_1(x) = x - 1/2$, $B_2(x) = x^2 - x + 1/6$, $B_3(x) = x^3 - 3x^2/2 + x/2$, $B_4(x) = x^4 - 2x^3 + x^2 - 1/30$. Furthermore, the Bernoulli numbers satisfy $B_m = B_m(0)$. They are accordingly given as coefficients of $z^m/m!$ in the series expansion of $z/(\exp(z) - 1)$. Thus,

$$\frac{z}{\exp(z) - 1} = \sum_{m=0}^{\infty} B_m\frac{z^m}{m!}, \quad |z| < 2\pi. \tag{5.62}$$

The first few Bernoulli numbers are $B_0 = 1$, $B_1 = -1/2$, $B_2 = 1/6$, $B_3 = 0$, $B_4 = -1/30$. All Bernoulli numbers of odd order exceeding one are equal to zero. This fact can be used to rewrite (5.62) in the following way:

$$\frac{z}{\exp(z) - 1} = 1 - \frac{z}{2} + \sum_{m=1}^{\infty} \frac{B_{2m}}{(2m)!}z^{2m}, \quad |z| < 2\pi. \tag{5.63}$$

An elementary derivation of the Euler-Maclaurin formula is given by Apostol [5]. We shall use the Euler-Maclaurin formula with $M = 1$. It can be written

$$\sum_{j=k}^{K} f(j) = \int_k^K f(x)\,dx + \frac{1}{2}[f(k) + f(K)] + \bar{R}_1, \quad k < K. \tag{5.64}$$

We derive an upper bound of the absolute value of the remainder term $\bar{R}_1$. Clearly, we have

$$\bar{R}_1 = \frac{1}{2}\int_k^K (B_2 - B_2(x - [x]))f''(x)\,dx. \tag{5.65}$$

Here, we show that the absolute value of the factor $B_2 - B_2(x - [x])$ is less than or equal to one. We put $u = x - [x]$, and note that $0 \le u \le 1$. We get therefore, for arbitrary real value of x,

$$B_2 - B_2(x - [x]) = B_2 - B_2(u) = \frac{1}{6} - \left(u^2 - u + \frac{1}{6}\right) = u - u^2, \quad 0 \le u \le 1. \tag{5.66}$$

Since the function $u - u^2$ takes values in the interval from 0 to 1/4 when u lies in the closed unit interval, we conclude that

$$|B_2 - B_2(x - [x])| \le \frac{1}{4}, \quad x \text{ real}. \tag{5.67}$$

This leads to the following inequality for the absolute value of the remainder term $\bar{R}_1$:

$$|\bar{R}_1| < \frac{1}{8}\int_k^K |f''(x)|\,dx. \tag{5.68}$$

To take the first step in the derivation of (5.55), we put $f(j) = \varphi(\hat{y}(j))/\sigma$, and $k = -\infty$, and choose K such that $\hat{z}(K) = o(1)$ and $\hat{y}(K) < -1$, and use the Euler-Maclaurin formula with $M = 1$, given in (5.64), with $|\bar{R}_1|$ bounded by (5.68).

The integral in the Euler-Maclaurin formula is then equal to

$$\int_{-\infty}^{K} f(x)\,dx = \frac{1}{\sigma}\int_{-\infty}^{K} \varphi(\hat{y}(x))\,dx = \int_{-\infty}^{\hat{y}(K)} \varphi(y)\,dy = \Phi(\hat{y}(K)), \qquad (5.69)$$

while the average of the boundary values is

$$\frac{1}{2}[f(-\infty) + f(K)] = \frac{1}{2\sigma}\varphi(\hat{y}(K)). \qquad (5.70)$$

To determine the upper bound of the absolute value of the remainder term $\bar{R}_1$, we evaluate the second derivative of f:

$$f''(x) = \frac{1}{\sigma^3}\varphi''(\hat{y}(x)) = \frac{1}{\sigma^3}He_2(\hat{y}(x))\varphi(\hat{y}(x)) = \frac{1}{\sigma^3}(\hat{y}^2(x) - 1)\varphi(\hat{y}(x)). \qquad (5.71)$$

Here He_n is used to denote a Hermite polynomial. The upper bound of the absolute value of the remainder term $\bar{R}_1$ is therefore found to be

$$|\bar{R}_1| \leq \frac{1}{8\sigma^3}\int_{-\infty}^{K} |\hat{y}^2(x) - 1|\varphi(\hat{y}(x))\,dx = \frac{1}{8\sigma^2}\int_{-\infty}^{\hat{y}(K)} |y^2 - 1|\varphi(y)\,dy$$

$$= \frac{1}{8\sigma^2}\int_{-\infty}^{\hat{y}(K)} (y^2 - 1)\varphi(y)\,dy,$$

$$\hat{z}(K) = o(1), \quad \hat{y}(K) < -1, \quad \sigma \to \infty. \qquad (5.72)$$

The removal of the absolute value signs around the polynomial $y^2 - 1$ is allowed since $y < -1$ throughout the integrand, which implies that $y^2 - 1 > 0$. Integration is easy, since the last integrand is equal to $\varphi''(y)$. Thus, we get

$$|\bar{R}_1| \leq \frac{1}{8\sigma^2}\varphi'(\hat{y}(K)) = -\frac{1}{8\sigma^2}\hat{y}(K)\varphi(\hat{y}(K)) = -\frac{1}{8\sigma}\hat{z}(K)\varphi(\hat{y}(K)),$$

$$\hat{z}(K) = o(1), \quad \hat{y}(K) < -1, \quad \sigma \to \infty. \qquad (5.73)$$

This is of smaller order than the average of the boundary values since $\hat{z}(K) = o(1)$. We conclude therefore that (5.55) holds for those K-values for which $\hat{y}(K) < -1$.

Next, as a preparation for proving (5.55) for $\hat{y}(K) \geq -1$, we note from the definition of the function G_2 in (5.19) that

$$G_2(x,z) \sim x, \quad xz \to 0. \qquad (5.74)$$

We conclude in particular that

$$G_2(x, \hat{z}(k)) \sim x, \quad x = O(1), \quad \hat{z}(k) = o(1), \quad \sigma \to \infty. \tag{5.75}$$

By applying this approximation of G_2 in the asymptotic approximation in (5.27) with $x = \pm 1/2$ we get

$$\Phi\left(\hat{y}\left(k + \frac{1}{2}\right)\right) \sim \Phi(\hat{y}(k)) + \frac{1}{2\sigma}\varphi(\hat{y}(k)), \quad \hat{z}(k) = o(1), \quad \sigma \to \infty, \tag{5.76}$$

$$\Phi\left(\hat{y}\left(k - \frac{1}{2}\right)\right) \sim \Phi(\hat{y}(k)) - \frac{1}{2\sigma}\varphi(\hat{y}(k)), \quad \hat{z}(k) = o(1), \quad \sigma \to \infty, \tag{5.77}$$

which implies that

$$\Phi\left(\hat{y}\left(k + \frac{1}{2}\right)\right) - \Phi\left(\hat{y}\left(k - \frac{1}{2}\right)\right) \sim \frac{1}{\sigma}\varphi(\hat{y}(k)), \quad \hat{z}(k) = o(1), \quad \sigma \to \infty. \tag{5.78}$$

Now let k_1 and K be chosen such that $\hat{y}(k_1) < -1$ and $\hat{y}(K) \geq -1$, and $\hat{z}(k_1) = o(1)$ and $\hat{z}(K) = o(1)$ as $\sigma \to \infty$. By applying the approximation in (5.76) we find from (5.55) that

$$\frac{1}{\sigma}\sum_{j=-\infty}^{k_1}\varphi(\hat{y}(j)) \sim \Phi\left(\hat{y}\left(k_1 + \frac{1}{2}\right)\right), \quad \hat{y}(k_1) < -1, \quad \hat{z}(k_1) = o(1), \quad \sigma \to \infty. \tag{5.79}$$

Furthermore, repeated application of (5.78) gives

$$\frac{1}{\sigma}\sum_{j=k_1+1}^{K}\varphi(\hat{y}(j)) \sim \Phi\left(\hat{y}\left(K + \frac{1}{2}\right)\right) - \Phi\left(\hat{y}\left(k_1 + \frac{1}{2}\right)\right),$$

$$\hat{y}(k_1) < -1, \quad \hat{z}(k_1) = o(1), \quad \hat{y}(K) \geq -1, \quad \hat{z}(K) = o(1), \quad \sigma \to \infty. \tag{5.80}$$

Combining the last two results, we get

$$\frac{1}{\sigma}\sum_{j=-\infty}^{K}\varphi(\hat{y}(j)) \sim \Phi\left(\hat{y}\left(K + \frac{1}{2}\right)\right), \quad \hat{y}(K) \geq -1, \quad \hat{z}(K) = o(1), \quad \sigma \to \infty. \tag{5.81}$$

By again using the approximation (5.76), we conclude that the approximation claimed in (5.55) holds also for K-values such that $\hat{y}(K) \geq -1$. This concludes the derivation of (5.55).

Additional terms in asymptotic approximations of $\sum_{j=-\infty}^{K}\varphi(\hat{y}(j)/\sigma$ for $\hat{z}(K) = o(1)$ can be derived in a similar way by increasing the value of M. Each added term is then of lower order than the preceding one. We comment that the conjectured asymptotic expansion given in (5.50) is based on the Euler-Maclaurin formula. However, in case $\hat{z}(K)$ is of the order of one, then we encounter the complication that all M-values give contributions of the same order.

We proceed to establish the approximation given in (5.56). To this end, we take K such that $\hat{z}(K) > 0$ and $\hat{z}(K)$ is of order one or larger. This means that K lies in the right far-end tail of the distribution with mean μ and standard deviation σ. Furthermore, we let m be a positive integer. By using (5.17) and (5.15) we get the following bound and asymptotic approximation of $\varphi(\hat{y}(K+m))$:

$$\varphi(\hat{y}(K+m)) < \varphi(\hat{y}(K))\exp(-m\hat{z}(K)),$$

$$\hat{z}(K) > 0, \quad \hat{z}(K) \text{ of order of 1 or larger}, \quad \sigma \to \infty, \qquad (5.82)$$

$$\varphi(\hat{y}(K+m)) \sim \varphi(\hat{y}(K))\exp(-m\hat{z}(K)),$$

$$\hat{z}(K) > 0, \quad \hat{z}(K) \text{ of order of 1 or larger}, \quad m = o(\sigma), \quad \sigma \to \infty. \quad (5.83)$$

By using the bound in (5.82) we get

$$\frac{1}{\sigma}\sum_{j=K}^{\infty}\varphi(\hat{y}(j)) < \frac{1}{\sigma}\varphi(\hat{y}(K))[1 + \exp(-\hat{z}(K)) + \exp(-2\hat{z}(K)) + \ldots]$$

$$= \frac{1}{\sigma}\varphi(\hat{y}(K))\frac{1}{1 - \exp(-\hat{z}(K))},$$

$$\hat{z}(K) > 0, \quad \hat{z}(K) \text{ of order of 1 or larger}, \quad \sigma \to \infty. \quad (5.84)$$

On the other hand, we use the asymptotic approximation in (5.83) to derive

$$\frac{1}{\sigma}\sum_{j=K}^{K+M-1}\varphi(\hat{y}(j))$$

$$\sim \frac{1}{\sigma}\varphi(\hat{y}(K))[1 + \exp(-\hat{z}(K)) + \exp(-2\hat{z}(K)) + \cdots + \exp(-(M-1)\hat{z}(K))]$$

$$= \frac{1}{\sigma}\varphi(\hat{y}(K))\frac{1 - \exp(-M\hat{z}(K))}{1 - \exp(-\hat{z}(K))},$$

$$M = o(\sigma), \quad \hat{z}(K) > 0, \quad \hat{z}(K) \text{ of order of 1 or larger}, \quad \sigma \to \infty. \quad (5.85)$$

This approximation can be simplified if M is chosen so large that $M\hat{z}(K) \to \infty$ as $\sigma \to \infty$. This is clearly possible for the given order of $\hat{z}(K)$, for example by taking $M = \left[\sqrt{\sigma}\right]$. Thus, we get

$$\frac{1}{\sigma}\sum_{j=K}^{K+M-1}\varphi(\hat{y}(j)) \sim \frac{1}{\sigma}\varphi(\hat{y}(K))\frac{1}{1 - \exp(-\hat{z}(K))},$$

$$M = \left[\sqrt{\sigma}\right], \quad \hat{z}(K) > 0, \quad \hat{z}(K) \text{ of order 1 or larger}, \quad \sigma \to \infty. \quad (5.86)$$

Since clearly $\sum_{j=K}^{\infty} \varphi(\hat{y}(j)) > \sum_{j=K}^{K+M-1} \varphi(\hat{y}(j))$, we conclude from (5.84) and (5.86) that the following asymptotic approximation holds:

$$\frac{1}{\sigma} \sum_{j=K}^{\infty} \varphi(\hat{y}(j)) \sim \frac{1}{\sigma} \varphi(\hat{y}(K)) \frac{1}{1 - \exp(-\hat{z}(K))},$$

$$\hat{z}(K) > 0, \quad \hat{z}(K) \text{ of order 1 or larger}, \quad \sigma \to \infty. \qquad (5.87)$$

In order to proceed we note next that the sum $\sum_{j=-\infty}^{\infty} \varphi(\hat{y}(j))/\sigma$ deviates from the value one by an exponentially small amount as $\sigma \to \infty$:

$$\frac{1}{\sigma} \sum_{j=-\infty}^{\infty} \varphi(\hat{y}(j)) \sim 1, \quad \sigma \to \infty. \qquad (5.88)$$

We establish the validity of this approximation at the end of the section. By using this result, we can write

$$\frac{1}{\sigma} \sum_{j=-\infty}^{K} \varphi(\hat{y}(j)) \sim 1 - \frac{1}{\sigma} \sum_{j=K}^{\infty} \varphi(\hat{y}(j)) + \frac{1}{\sigma} \varphi(\hat{y}(K)), \quad \sigma \to \infty. \qquad (5.89)$$

Applying the approximation in (5.87), we get

$$\frac{1}{\sigma} \sum_{j=-\infty}^{K} \varphi(\hat{y}(j)) \sim 1 + \frac{1}{\sigma} \varphi(\hat{y}(K)) \left[1 - \frac{1}{1 - \exp(-\hat{z}(K))} \right]$$

$$= 1 - \frac{1}{\sigma} \varphi(\hat{y}(K)) \frac{1}{\exp(\hat{z}(K)) - 1},$$

$$\hat{z}(K) > 0, \quad \hat{z}(K) \text{ of order 1 or larger}, \quad \sigma \to \infty. \qquad (5.90)$$

This shows that (5.56) holds.

The derivation of (5.57) is similar, but slightly shorter. We take K such that $\hat{z}(K) < 0$ and $\hat{z}(K)$ is of the order of one or larger. We can then apply the inequality (5.17) to get the following upper bound of the sum $\sum_{j=-\infty}^{K} \varphi(\hat{y}(j))/\sigma$:

$$\frac{1}{\sigma} \sum_{j=-\infty}^{K} \varphi(\hat{y}(j)) < \frac{1}{\sigma} \varphi(\hat{y}(K))[1 + \exp(\hat{z}(K)) + \exp(2\hat{z}(K)) + \ldots]$$

$$= \frac{1}{\sigma} \varphi(\hat{y}(K)) \frac{1}{1 - \exp(\hat{z}(K))},$$

$$\hat{z}(K) < 0, \quad \hat{z}(K) \text{ of order of 1 or larger}. \qquad (5.91)$$

A lower bound is easily found by omitting all but finitely many terms of the sum over j:

$$\frac{1}{\sigma} \sum_{j=-\infty}^{K} \varphi(\hat{y}(j)) > \frac{1}{\sigma} \sum_{j=K-M+1}^{K} \varphi(\hat{y}(j)), \quad M > 1. \tag{5.92}$$

The terms in the sum over j in this lower bound can be approximated by the aid of (5.15). We get

$$\frac{1}{\sigma} \sum_{j=K-M+1}^{K} \varphi(\hat{y}(j)) \sim \frac{1}{\sigma} \varphi(\hat{y}(K))[1 + \exp(\hat{z}(K)) + \exp(2\hat{z}(K)) + \cdots + \exp((M-1)\hat{z}(K))]$$

$$= \frac{1}{\sigma} \varphi(\hat{y}(K)) \frac{1 - \exp(M\hat{z}(K))}{1 - \exp(\hat{z}(K))} \sim \frac{1}{\sigma} \varphi(\hat{y}(K)) \frac{1}{1 - \exp(\hat{z}(K))},$$

$$M = \left[\sqrt{\sigma}\right], \quad \hat{z}(K) < 0, \quad \hat{z}(K) \text{ of order of 1 or larger,} \quad \sigma \to \infty. \tag{5.93}$$

We conclude that the asymptotic approximation of the lower bound of the sum that we are concerned with is equal to the upper bound. It follows that the asymptotic approximation of the sum is equal to this common value. Thus, (5.57) has been derived.

We indicate the arguments necessary to conclude that (5.88) holds. Indeed, we claim the following more general result:

$$\frac{1}{\sigma} \sum_{j=-\infty}^{\infty} \varphi(\hat{y}(j)) = 1 + 2 \sum_{k=1}^{\infty} \exp(-2k^2\pi^2\sigma^2) \cos(2\pi k\mu). \tag{5.94}$$

To derive it, we use the Poisson formula in the form

$$\sum_{j=-\infty}^{\infty} f(j-\mu) = \lim_{n \to \infty} \sum_{k=-N}^{N} \exp(-2\pi ki\mu) \int_{-\infty}^{\infty} \exp(-2\pi kiz) f(z) \, dz, \tag{5.95}$$

as given by de Bruijn [22], and substitute $f(k) = \varphi(k/\sigma)/\sigma$, which implies $f(j - \mu) = \varphi(\hat{y}(j))/\sigma$. With this f, the integral is found to be equal to $\exp(-2k^2\pi^2\sigma^2)$. Insertion into (5.95) leads to the explicit expression in (5.94).

5.5   An Approximation of a Sum of Reciprocals of Normal Densities

We give here an approximation of a sum of reciprocal normal densities of the form $\frac{1}{\sigma} \sum_{j=k+1}^{K} \varphi^*(\hat{y}(j))$. This differs from the sum of normal densities considered in the previous section, where the summation over j went from $-\infty$ to K. It is not possible to start the summation at $j = -\infty$ for the reciprocal normal density, since the function $\varphi^*(y)$ is only defined for finite y.

The main result is stated as a theorem that gives an asymptotic approximation of the sum of reciprocal normal densities. We cite also a conjectured asymptotic expansion, and finally we spell out the results of applying the theorem in two particular cases where the approximation takes different forms.

Our main result is given by the following theorem:

Theorem 5.5 *The sum of reciprocal normal densities* $\frac{1}{\sigma}\sum_{j=k+1}^{K}\varphi^*(\hat{y}(j))$ *has the following asymptotic approximation:*

$$\frac{1}{\sigma}\sum_{j=k+1}^{K}\varphi^*(\hat{y}(j)) \sim F^*(K) - F^*(k), \quad k < K, \quad \sigma \to \infty, \tag{5.96}$$

where

$$F^*(j) = \Phi^*(\hat{y}(j)) + \frac{1}{\sigma}\varphi^*(\hat{y}(j))G_1(\hat{z}(j)). \tag{5.97}$$

We add the conjecture that the same sum of reciprocal normal densities has an asymptotic expansion that is expressed by (5.96), with the function F^* given by

$$F^*(j) = \Phi^*(\hat{y}(j)) + \frac{1}{\sigma}\varphi^*(\hat{y}(j)) \sum_{m=0}^{\infty}\frac{1}{m!}\left(\frac{1}{2\sigma^2}\right)^m G_1^{(2m)}(\hat{z}(j)). \tag{5.98}$$

We leave it as an open problem to derive this result.

We use the result in Theorem 5.5 to derive two asymptotic approximations that take different forms, depending on the values of K or k. One case applies when K or k lies in the body or a near-end tail of the distribution with mean μ and standard deviation σ, while another case arises when K or k lie in a far-end tail.

The results are contained in the following corollary:

Corollary 5.5 *The sum of reciprocal normal densities* $\frac{1}{\sigma}\sum_{j=k+1}^{K}\varphi^*(\hat{y}(j))$ *has the asymptotic approximation that is expressed by (5.96), with the function F^* given by*

$$F^*(j) \sim \Phi^*(\hat{y}(j)) + \frac{1}{2\sigma}\varphi^*(\hat{y}(j)), \quad \text{if } \hat{z}(j) = o(1), \quad \sigma \to \infty, \tag{5.99}$$

or

$$F^*(j) \sim \frac{1}{\sigma}\varphi^*(\hat{y}(j))\frac{1}{1-\exp(-\hat{z}(j))}, \quad \text{if } \hat{z}(j) \text{ is of order 1 or larger}, \quad \sigma \to \infty. \tag{5.100}$$

Proof. The result in (5.99) follows from (5.97), since $\hat{z}(j) = o(1)$ implies that $G_1(\hat{z}(j)) \sim 1/2$.

To derive (5.100) we note first that $|\hat{y}(j)| \to \infty$. Therefore we get the following asymptotic approximation from Corollary 5.3: $\Phi^*(y) \sim \varphi^*(y)/y$ as $|y| \to \infty$. This leads to the following approximation of $F^*(j)$:

$$F^*(j) \sim \frac{\varphi^*(\hat{y}(j))}{\hat{y}(j)} + \frac{1}{\sigma}\varphi^*(\hat{y}(j)G_1(\hat{z}(j))) = \frac{1}{\sigma}\varphi^*(\hat{y}(j))\left(\frac{1}{\hat{z}(j)} + G_1(\hat{z}(j))\right),$$

$$\hat{z}(j) \text{ of order 1 or larger}, \quad \sigma \to \infty. \quad (5.101)$$

By using the definition of the function G_1 in (5.48) we find that this approximation is equivalent to the one given in (5.100). $\square$

We turn now toward a derivation of the theorem given above.

We derive first the approximation (5.96) with F^* given by (5.99) in case both k and K lie in the body or a near-end tail of the distribution with mean μ and standard deviation σ. Thus, we have $\hat{z}(K) = o(1)$ and $\hat{z}(k) = o(1)$, and also $\hat{z}(j) = o(1)$ for $k < j < K$. We apply the asymptotic approximation of $\Phi^*(\hat{y}(j+x))$ given in (5.28), with the two x-values $1/2$ and $-1/2$. We get then

$$\Phi^*(\hat{y}(j+1/2)) \sim \Phi^*(\hat{y}(j)) + \frac{1}{2\sigma}\varphi^*(\hat{y}(j)), \quad \hat{z}(j) = o(1), \quad \sigma \to \infty, \quad (5.102)$$

$$\Phi^*(\hat{y}(j-1/2)) \sim \Phi^*(\hat{y}(j)) - \frac{1}{2\sigma}\varphi^*(\hat{y}(j)), \quad \hat{z}(j) = o(1), \quad \sigma \to \infty. \quad (5.103)$$

This implies that

$$\Phi^*(\hat{y}(j+1/2)) - \Phi^*(\hat{y}(j-1/2)) \sim \frac{1}{\sigma}\varphi^*(\hat{y}(j)), \quad \hat{z}(j) = o(1), \quad \sigma \to \infty. \quad (5.104)$$

By applying this approximation in the sum $\sum_{j=k+1}^{K} \varphi^*(\hat{y}(j))/\sigma$ we get

$$\frac{1}{\sigma}\sum_{j=k+1}^{K} \varphi^*(\hat{y}(j)) \sim \Phi^*(\hat{y}(K+1/2)) - \Phi^*(\hat{y}(k+1/2)),$$

$$k < K, \quad \hat{z}(k) = o(1), \quad \hat{z}(K) = o(1), \quad \sigma \to \infty. \quad (5.105)$$

Here, it is straightforward to use the approximation (5.102) to show that (5.96) holds with F^* given by (5.99).

Next we consider the situation where k is in the body or in a near-end tail, while K is in the right far-end tail of a distribution with mean μ and standard deviation σ. Thus, we have $\hat{z}(k) = o(1)$ and $k < K$, while $\hat{z}(K)$ is of the order of one or larger. In this case we note first that $\varphi^*(\hat{y}(j))$ is an increasing function of j whenever $j > \mu$ so that $\hat{y}(j)$ is positive. Thus, the largest contribution to the sum $\sum_{j=k+1}^{K} \varphi^*(\hat{y}(j))$ comes from the largest j-values. We determine therefore an asymptotic approximation of the sum $\sum_{j=K-M+1}^{K} \varphi^*(\hat{y}(j))$ for a suitable value of M, and show that the remainder of the sum, $\sum_{j=k+1}^{K-M} \varphi^*(\hat{y}(j))$, can be ignored asymptotically. We choose the positive integer M so that $M \to \infty$ as $\sigma \to \infty$, but $M = o(\sigma)$. To be specific, we take $M = \lceil\sqrt{\sigma}\rceil$. It follows then that all integers in the interval from $-M$ to -1 are of smaller order then σ. We can therefore apply the asymptotic approximation in (5.16) to get

$$\varphi^*(\hat{y}(K-i)) \sim \varphi^*(\hat{y}(K)) \exp(-i\hat{z}(K)),$$

$$-M \le i \le -1, \quad \hat{z}(K) \text{ of the order of 1 or larger}, \quad \sigma \to \infty. \quad (5.106)$$

By inserting these approximations into the sum $\sum_{j=K-M+1}^{K} \varphi^*(\hat{y}(j))/\sigma$, we get

$$\frac{1}{\sigma} \sum_{j=K-M+1}^{K} \varphi^*(\hat{y}(j))$$

$$\sim \frac{1}{\sigma} \varphi^*(\hat{y}(K)) [1 + \exp(-\hat{z}(K)) + \exp(-2\hat{z}(K)) + \cdots + \exp(-(M-1)\hat{z}(K))]$$

$$= \frac{1}{\sigma} \varphi^*(\hat{y}(K)) \frac{1 - \exp(-M\hat{z}(K))}{1 - \exp(-\hat{z}(K))} \sim \frac{1}{\sigma} \varphi^*(\hat{y}(K)) \frac{1}{1 - \exp(-\hat{z}(K))},$$

$$\hat{z}(K) > 0, \quad \hat{z}(K) \text{ of order of 1 or larger}, \quad M = \left[\sqrt{\sigma}\right], \quad \sigma \to \infty. \quad (5.107)$$

A comparison with (5.100) with $j = K$ shows that this agrees with the approximation for $F^*(K)$ claimed in this case. It remains to show that both the remainder of the sum, $\sum_{j=k+1}^{K-M} \varphi^*(\hat{y}(j))/\sigma$, and $F^*(k)$ can be ignored asymptotically.

A course upper bound for the remainder of the sum is found by replacing each term by its maximum value $\varphi^*(\hat{y}(K-M))$. In addition, the number of terms is certainly bounded by $2K$. Thus we get

$$\frac{1}{\sigma} \sum_{j=k+1}^{K-M} \varphi^*(\hat{y}(j)) < \frac{1}{\sigma} 2K \varphi^*(\hat{y}(K-M)) \sim \frac{1}{\sigma} \varphi^*(\hat{y}(K)) 2K \exp(-M\hat{z}(K)),$$

$$k + 1 \ge \mu, \quad \hat{z}(k) = o(1), \quad k < K, \quad \hat{z}(K) \text{ of the order of 1 or larger},$$

$$M = \left[\sqrt{\sigma}\right], \quad \sigma \to \infty, \quad (5.108)$$

where we have used the approximation (5.16) of $\varphi^*(\hat{y}(K-M))$. Here, the factor $K \exp(-M\hat{z}(K))$ is exponentially small. This proves that the remainder of the sum can be ignored asymptotically.

Finally, we show that $F^*(k)$ in this case can be ignored asymptotically. This follows from the fact that

$$\varphi^*(\hat{y}(k))/\varphi^*(\hat{y}(K)) = \exp\left(\frac{1}{2}\hat{y}^2(k) - \frac{1}{2}\hat{y}^2(K)\right) = \exp\left(-\frac{\sigma^2}{2}\left(\hat{z}^2(K) - \hat{z}^2(k)\right)\right).$$

$$(5.109)$$

This ratio is exponentially small as $\sigma \to \infty$ in the case treated here, with $\hat{z}(K)$ of the order of one or larger, and $k < K$, and $\hat{z}(k) = o(1)$. This concludes the derivation of the asymptotic approximation (5.96) in this case.

There are other cases to consider. One is when both k and K lie in the right far-end tail of the distribution with mean μ and standard deviation σ. The validity of (5.96) follows then by taking k_0 as an integer with $\hat{z}(k_0) = o(1)$, and writing

$$\frac{1}{\sigma}\sum_{j=k+1}^{K}\varphi^*(\hat{y}(j)) = \frac{1}{\sigma}\sum_{j=k_0+1}^{K}\varphi^*(\hat{y}(j)) - \frac{1}{\sigma}\sum_{j=k_0+1}^{k}\varphi^*(\hat{y}(j)), \tag{5.110}$$

and applying the result just established to each of these two terms.

Another case occurs when k lies in the left far-end tail, while K lies in the body or a near-end tail of the distribution with mean μ and standard deviation σ. By applying (5.16) we get

$$\frac{1}{\sigma}\sum_{j=k+1}^{k+M}\varphi^*(\hat{y}(j))$$

$$\sim \frac{1}{\sigma}\varphi^*(\hat{y}(k))\exp(\hat{z}(k))\left[1+\exp(\hat{z}(k))+\exp(2\hat{z}(k))+\cdots+\exp((M-1)\hat{z}(k))\right]$$

$$= \frac{1}{\sigma}\varphi^*(\hat{y}(k))\exp(\hat{z}(k))\frac{1-\exp(M\hat{z}(k))}{1-\exp(k\hat{z}(k))} \sim \frac{1}{\sigma}\varphi^*(\hat{y}(k))\frac{\exp(\hat{z}(k))}{1-\exp(\hat{z}(k))}$$

$$= -\frac{1}{\sigma}\varphi^*(\hat{y}(k))\frac{1}{1-\exp(-\hat{z}(k))},$$

$$\hat{z}(k)<0, \quad \hat{z}(k) \text{ of order of 1 or larger}, \quad M=\left[\sqrt{\sigma}\right], \quad \sigma\to\infty. \tag{5.111}$$

A comparison with (5.100) shows that this agrees with the approximation of $-F^*(k)$ in this case. We note that this approximation is positive, since $F^*(k)$ is negative. In similarity to the case treated above, we show that the remainder of the sum, $\sum_{j=k+M+1}^{K}\varphi^*(\hat{y}(j))/\sigma$, and $F^*(K)$ can be ignored asymptotically. Thus, the approximation in (5.96) holds in this case.

The case when both k and K lie in the left far-end tail of the distribution with mean μ and standard deviation σ is treated as above by taking k_0 in the body or a near-end tail of the same distribution, and writing

$$\sum_{j=k+1}^{K}\varphi^*(\hat{y}(j)) = \sum_{j=k+1}^{k_0}\varphi^*(\hat{y}(j)) - \sum_{j=K+1}^{k_0}\varphi^*(\hat{y}(j)), \quad k<K, \tag{5.112}$$

and applying the result just derived to both of these sums.

Finally, we consider the case when k lies in the left far-end tail and K lies in the right far-end tail of the distribution with mean μ and standard deviation σ. We then take k_0 in the body or a near-end tail of the same distribution, and write

$$\sum_{j=k+1}^{K}\varphi^*(\hat{y}(j)) = \sum_{j=k+1}^{k_0}\varphi^*(\hat{y}(j)) + \sum_{j=k_0+1}^{K}\varphi^*(\hat{y}(j)). \tag{5.113}$$

The approximation (5.96) of $\sum_{j=k+1}^{K} \varphi^*(\hat{y}(j))/\sigma$ follows by applying the results derived above to both of the terms in the right-hand side of this expression.

5.6   On the Cox Continuity Correction

An alternative way of dealing with a sum of normal densities is via the so-called continuity correction $C(k)$ defined by Cox [20] as follows:

$$\frac{1}{\sigma} \sum_{j=-\infty}^{K} \varphi(\hat{y}(j)) = \Phi(\hat{y}(K+C(K))). \qquad (5.114)$$

We note that the continuity correction $C(K)$ depends on the values of the mean μ and the standard deviation σ, but that this dependence is not made explicit by the notation. The main result in this section is an asymptotic approximation of the Cox continuity correction. It can be written

$$C(K) \sim G(\hat{z}(K)), \quad \sigma \to \infty, \qquad (5.115)$$

where the function G is defined by

$$G(z) = \begin{cases} \dfrac{1}{z} \log \dfrac{\exp(z) - 1}{z}, & z \neq 0, \\[2mm] \dfrac{1}{2}, & z = 0. \end{cases} \qquad (5.116)$$

A plot of the function G is given in Fig. 5.2. This figure also shows the function G_1 defined by (5.48). It is straightforward to show that G, like G_1, takes values in the open unit interval $(0,1)$, is monotonically increasing, approaches zero and one as the argument z approaches $\pm\infty$, and that $G(z) + G(-z) = 1$.

To derive the above approximation of the Cox continuity correction, we approximate the left-hand side of the Cox definition (5.114) by (5.49), and the right-hand side by (5.27). The first of these two approximations gives

$$\frac{1}{\sigma} \sum_{j=-\infty}^{K} \varphi(\hat{y}(j)) \sim \Phi(\hat{y}(K)) + \frac{1}{\sigma} \varphi(\hat{y}(K))[1 - G_1(\hat{z}(K))], \quad \sigma \to \infty, \qquad (5.117)$$

while the second one can be written

$$\Phi(\hat{y}(K+C(K))) \sim \Phi(\hat{y}(K)) + \frac{1}{\sigma} \varphi(\hat{y}(K)) G_2(C(K), \hat{z}(K)), \quad \sigma \to \infty. \qquad (5.118)$$

Equality between the two left-hand sides gives an asymptotic relation between the two right-hand sides:

$$G_2(C(K), \hat{z}(K)) \sim 1 - G_1(\hat{z}(K)), \quad \sigma \to \infty. \tag{5.119}$$

By using the definitions of the the functions G_1 and G_2 we get

$$\exp(\hat{z}(K)C(K)) \sim \frac{\exp(\hat{z}(K)) - 1}{\hat{z}(K)}, \quad \sigma \to \infty. \tag{5.120}$$

We note next from the definition of G that the following relation holds:

$$\exp(zG(z)) = \frac{\exp(z) - 1}{z}. \tag{5.121}$$

After evaluating both sides of this relation at $z = \hat{z}(K)$, we find that

$$\exp(\hat{z}(K)C(K)) \sim \exp(\hat{z}(K)G(\hat{z}(K))), \quad \sigma \to \infty. \tag{5.122}$$

In case $\hat{z}(K) = o(1)$ we get $G(\hat{z}(K)) \sim 1/2$. This implies $C(K) \sim 1/2$. Thus $C(K)$ satisfies $C(K) \sim G(\hat{z}(K))$ as $\sigma \to \infty$. In case $\hat{z}(K)$ is of the order of one or larger, we get $\exp(\hat{z}(K)(C(K) - G(\hat{z}(K)))) \sim 1$, which gives $\hat{z}(K)(C(K) - G(\hat{z}(K))) = o(1)$ or $C(K) - G(\hat{z}(K)) = o(1)$. This implies $C(K)/G(\hat{z}(K)) - 1 = o(1)/G(\hat{z}(K)) = o(1)$. We conclude that $C(k)/G(\hat{z}(K)) \to 1$ as $\sigma \to \infty$, which is equivalent to $C(K) \sim G(\hat{z}(K))$. Thus, the asymptotic approximation $C(K) \sim G(\hat{z}(K))$ holds in either case.

Cox derives the approximation $C(K) \approx 1/2$, without restrictions on K. His derivation is based on the assumption that $\varphi(\hat{y}(k + x))$ is linear in x. Our discussion in Sect. 5.1 shows that this assumption holds (asymptotically) only in the body and the near-end tails, but not in the far-end tails of the distribution with mean μ and standard deviation σ. Thus, we conclude that the Cox approximation does not hold in the far-end tails of this distribution.

The Cox continuity correction is mainly used in statistical data analysis, where the data lies in the body of the distribution concerned. In that case, the approximation $C(K) \sim 1/2, \sigma \to \infty$, is supported by our result.

Chapter 6
Preparations for the Study of the Stationary Distribution $\mathbf{p}^{(1)}$ of the SIS Model

We devote the present chapter to derivations of results that are needed in the next chapter, where we deal with approximations of the stationary distribution $\mathbf{p}^{(1)}$ of the auxiliary process $\{X^{(1)}(t)\}$ of the SIS model. We recall from (3.8) that the stationary distribution $p_n^{(1)}$ is determined from the sequence ρ_n by

$$p_n^{(1)} = \frac{\rho_n}{\sum_{n=1}^{N} \rho_n}, \quad n = 1, 2, \ldots, N. \tag{6.1}$$

The fact that ρ_n is known explicitly and exactly by (4.6) in terms of the two essential parameters R_0 and N of the model and the value of n does not suffice for our purposes. Approximations of ρ_n are essential for progress. We deal with approximations of the numerator ρ_n of the right-hand side of the expression (6.1) for $p_n^{(1)}$ in the first two sections of this chapter, and with approximations of the denominator $\sum_{n=1}^{N} \rho_n$ in Sect. 6.3 and its three subsections. The approximations of ρ_n are rather direct consequences of the application of Stirling's formula, while the approximations of $\sum_{n=1}^{N} \rho_n$ are far more intricate. They are based on the approximation of the sum of normal densities given in Sect. 5.4.

6.1 A General Approximation of ρ_n

The exact expression for the sequence ρ_n given by (4.6) is valid for any integer value of n in the whole interval of values from 1 to N. We shall derive an approximation that is valid in a shorter interval. Our approximation takes the form of some constant multiplying a normal density with known mean and standard deviation. One important advantage with this approximation compared with the exact expression is that it allows us to evaluate an approximation of a sum of values of ρ_n over the interval of validity. This evaluation makes use of the approximations of sums of normal densities derived in Chap. 5. Important applications are reported

I. Nåsell, *Extinction and Quasi-stationarity in the Stochastic Logistic SIS Model*,
Lecture Notes in Mathematics 2022, DOI 10.1007/978-3-642-20530-9_6,
© Springer-Verlag Berlin Heidelberg 2011

in Sect. 6.3, where we actually give approximations of the sum $\sum_{n=1}^{N} \rho_n$ over all n-values in the whole interval from one to N. Derivations of these results in each of three parameter regions are found separately in Sects. 6.3.1–6.3.3.

We shall allow n to be real-valued, although the expression in (4.6) is valid only when n is an integer. The extension to real n-values is accomplished by replacing $(N-n)!$ in (4.6) by $\Gamma(N-n+1)$.

We introduce notation that will be used in the statement of our general approximation result. We let α be a positive quantity that takes values less than or equal to R_0, i.e. $0 < \alpha \leq R_0$. The following quantities are then defined:

$$m_a = \frac{R_0 - \alpha}{R_0} N, \qquad m_b = \frac{R_0 - A_1(\alpha)}{R_0} N, \qquad (6.2)$$

$$s_a = \sqrt{\frac{\alpha N}{R_0}}, \qquad s_b = s_a = \sqrt{\frac{\alpha N}{R_0}}, \qquad (6.3)$$

$$y_a(n) = \frac{n - m_a}{s_a}, \qquad y_b(n) = \frac{n - m_b}{s_b}, \qquad (6.4)$$

with the function A_1 defined by

$$A_1(\alpha) = \alpha(1 - \log \alpha), \qquad \alpha > 0. \qquad (6.5)$$

We note that the difference between m_b and m_a can be expressed as follows:

$$m_b - m_a = \alpha \log(\alpha) \frac{N}{R_0}. \qquad (6.6)$$

Furthermore, we define

$$h_0 = N \left[\log R_0 - 1 + \frac{A_2(\alpha)}{R_0} \right], \qquad (6.7)$$

with the function A_2 given by

$$A_2(\alpha) = A_1(\alpha) + \frac{1}{2}\alpha(\log \alpha)^2 = \frac{\alpha}{2}\left(1 + (1 - \log \alpha)^2\right). \qquad (6.8)$$

The functions A_1 and A_2 are plotted in Fig. 6.1.

For each α in the interval $0 < \alpha \leq R_0$ we shall derive an approximation of ρ_n that is proportional to $\varphi(y_b(n))$, and that is valid in an interval of n-values that is described by the requirement $y_a(n) = o(\sqrt{N})$. This means that the approximation is valid in the body and the near-end tails of the distribution with mean m_a and standard deviation σ_a. The fact that the approximation holds not only in the body, but also in the near-end tails is of value later on in this chapter, where we determine approximations of $\sum_{n=1}^{N} \rho_n$. The derivation shows that the approximation can be interpreted as an expansion about the point m_a.

Fig. 6.1 The functions $A_1(\alpha)$ and $A_2(\alpha)$ are shown as functions of α

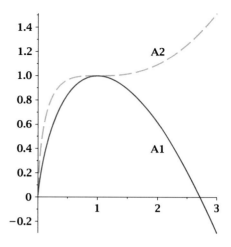

The result is given in the following theorem:

Theorem 6.1 *The quantity*

$$\rho_n = \frac{1}{R_0} \frac{N!}{(N-n)!} \left(\frac{R_0}{N}\right)^n, \quad n = 1, 2, \ldots, N, \tag{6.9}$$

is approximated as follows for $0 < \alpha \le R_0$:

$$\rho_n \sim \sqrt{\frac{2\pi}{\alpha R_0}} \exp(h_0)\varphi(y_b(n)), \quad y_a(n) = o(\sqrt{N}), \quad 1 \le n, \quad N \to \infty, \tag{6.10}$$

Proof. The starting point for the derivation of this approximation is the exact expression (4.6) for ρ_n. As mentioned above, we replace the factorials by Gamma-functions, using the relation $z! = \Gamma(z+1)$. The Gamma-functions are then approximated by using Stirling's formula: $\Gamma(z+1) \sim (z/e)^z \sqrt{2\pi z}$ as $z \to \infty$. The conditions for applying Stirling's formula are then satisfied if both $N \to \infty$ and $N - n \to \infty$. The resulting asymptotic approximation of ρ_n can be expressed as follows:

$$\rho_n \sim \frac{1}{R_0 \sqrt{1-n/N}} \frac{R_0^n}{(1-n/N)^{N-n} \exp(n)}, \quad n = 1, 2, \ldots, N, \quad N \to \infty, \quad N-n \to \infty. \tag{6.11}$$

This can be written

$$\rho_n \sim g(n)\exp(h(n)), \quad n = 1, 2, \ldots, N, \quad N \to \infty, \quad N-n \to \infty, \tag{6.12}$$

where

$$g(n) = \frac{1}{R_0} \frac{1}{\sqrt{1-n/N}}, \tag{6.13}$$

and

$$h(n) = n \log R_0 - n - (N-n) \log\left(1 - \frac{n}{N}\right), \tag{6.14}$$

The approximation of ρ_n given in (6.10) is based on Taylor expansions of $h(n)$ and $g(n)$ about the n-value m_a. As mentioned above, we avoid the α-value zero. There are two reasons for this. One is that $\alpha = 0$ would give $m_a = N$, and neither $h(n)$ nor $g(n)$ possess Taylor expansions about this point. A second reason is that the approximation (6.12) is not valid at $n = N$. We include three terms in the Taylor expansion of $h(n)$ and one term in the Taylor expansion of $g(n)$. The reason for this will be apparent later, when we show that these terms provide asymptotic approximations of the respective functions under appropriate restrictions on n.

We use $T\{f\}(n)$ to denote a formal Taylor expansion of any function $f(n)$. At the outset we shall not worry about the convergence of the Taylor expansions that we encounter. The Taylor expansion of $h(n)$ about $n = m_a$ can be written

$$T\{h\}(n) = h(m_a) + h'(m_a)(n - m_a) + \sum_{k=2}^{\infty} \frac{h^{(k)}(m_a)}{k!}(n - m_a)^k, \quad 0 < \alpha \le R_0. \tag{6.15}$$

To evaluate the derivatives here, we note that

$$h'(n) = \log\left(\left(1 - \frac{n}{N}\right) R_0\right), \tag{6.16}$$

and that

$$h^{(k)}(n) = -\frac{(k-2)!}{N^{k-1}} \frac{1}{(1 - n/N)^{k-1}}, \quad k = 2, 3, \dots. \tag{6.17}$$

Evaluations at $n = m_a$ are aided by observing that $1 - m_a/N = \alpha/R_0$. Thus, we get

$$h'(m_a) = \log \alpha, \tag{6.18}$$

and

$$h^{(k)}(m_a) = -(k-2)!\left(\frac{R_0}{\alpha N}\right)^{k-1}, \quad k = 2, 3, \dots. \tag{6.19}$$

Hence, the Taylor expansion of $h(n)$ can be written:

$$T\{h\}(n) = h(m_a) + \log(\alpha)(n - m_a) - \sum_{k=2}^{\infty} \frac{1}{(k-1)k}\left(\frac{R_0}{\alpha N}\right)^{k-1}(n - m_a)^k$$

$$= h(m_a) + \log(\alpha)(n - m_a) - \sum_{k=2}^{\infty} \frac{1}{(k-1)k} \frac{1}{s_a^{k-2}}\left(\frac{n - m_a}{s_a}\right)^k. \tag{6.20}$$

The successive terms in the summation over k are of decreasing order in N if n is restricted to values for which the product of the last two factors remains of order smaller than one. This means that we may allow $y_a(n) = (n - m_a)/s_a$ to grow with N as long as the growth is slower than the growth of s_a. Since the latter is of order of

$\sqrt{N}$, we are led to the result that the Taylor expansion in (6.20) will converge, and that it forms an asymptotic expansion of $h(n)$ as $N \to \infty$ when $y_a(n) = o(\sqrt{N})$. This occurs when n is restricted to lie in the body or the near-end tails of the distribution with mean m_a and standard deviation s_a. We use this result to conclude that the first three terms of the Taylor expansion give the following asymptotic approximation of $h(n)$ as $N \to \infty$:

$$h(n) \sim h(m_a) + \log(\alpha)(n - m_a) - \frac{1}{2}y_a^2(n),$$

$$y_a(n) = o(\sqrt{N}), \quad 0 < \alpha \le R_0, \quad N \to \infty. \tag{6.21}$$

Additional terms in the approximation of $h(n)$ can be gotten by including additional terms from its Taylor expansion. But if one requires or desires additional terms in this approximation, then it is also necessary to include corresponding additional terms in the approximation of $g(n)$ below, and also in the Stirling approximations of the Gamma functions replacing the factorials in (4.6).

The last step in the derivation of an asymptotic approximation of $h(n)$ is to rewrite the approximation given in (6.1) by completing the square. The resulting approximation of $h(n)$ is as follows:

$$h(n) \sim h_0 - \frac{1}{2}y_b^2(n), \quad 0 < \alpha \le R_0, \quad y_a(n) = O(1), \quad N \to \infty, \tag{6.22}$$

with h_0 given by (6.7).

To show this, we insert the definition of m_a into the expression (6.14) for $h(n)$. By using the relation $1 - m_a/n = \alpha/R_0$ we get

$$h(m_a) = \frac{R_0 - \alpha}{R_0}N(\log R_0 - 1) - \frac{\alpha N}{R_0}(\log \alpha - \log R_0) = N(\log R_0 - 1) + \frac{N}{R_0}A_1(\alpha). \tag{6.23}$$

Therefore, we can write

$$h(m_a) - h_0 = -\frac{1}{2}\frac{\alpha N}{R_0}\log^2 \alpha = -\frac{1}{2}s_a^2 \log^2 \alpha. \tag{6.24}$$

Furthermore,

$$m_b - m_a = s_a^2 \log \alpha. \tag{6.25}$$

Using this expression, we get

$$\frac{1}{2}y_a^2(n) - \frac{1}{2}y_b^2(n) = \frac{(n - m_a)^2 - (n - m_b)^2}{2s_a^2}$$

$$= \frac{(n - m_a - (m_b - m_a)/2)(m_b - m_a)}{s_a^2}$$

$$= (n - m_a)\log \alpha - \frac{1}{2}s_a^2 \log^2 \alpha. \tag{6.26}$$

Subtraction from (6.24) leads to the relation

$$h(m_a) + (n - m_a) \log \alpha - \frac{1}{2} y_a^2(n) = h_0 - \frac{1}{2} y_b^2(n). \tag{6.27}$$

This establishes the approximation of $h(n)$ given in (6.22).

We proceed to deal with the function $g(n)$ in a similar way. Its Taylor expansion about $n = m_a$ can be written

$$T\{g\}(n) = \sum_{k=0}^{\infty} \frac{g^{(k)}(m_a)}{k!} (n - m_a)^k, \quad 0 < \alpha \le R_0. \tag{6.28}$$

The derivative of $g(n)$ of order k equals

$$g^{(k)}(n) = \frac{1}{R_0} \frac{(2k)!}{2^{2k} k!} \frac{1}{N^k} \frac{1}{(1 - n/N)^{k+1/2}}, \quad k = 0, 1, 2, \ldots. \tag{6.29}$$

Evaluating the derivative at $n = m_a$, we use again the relation $1 - m_a/N = \alpha/R_0$, to get

$$g^{(k)}(m_a) = \frac{1}{\sqrt{\alpha R_0}} \frac{(2k)!}{2^{2k} k!} \left(\frac{R_0}{\alpha N} \right)^k, \quad k = 0, 1, 2, \ldots. \tag{6.30}$$

Thus, the Taylor expansion of $g(n)$ can be written as follows:

$$T\{g\}(n) = \frac{1}{\sqrt{\alpha R_0}} \sum_{k=0}^{\infty} \frac{1}{2^{2k}} \binom{2k}{k} \left(\frac{R_0}{\alpha N} \right)^k (n - m_a)^k = \frac{1}{\sqrt{\alpha R_0}} \sum_{k=0}^{\infty} \frac{1}{2^{2k}} \binom{2k}{k} \frac{1}{s_a^k} y_a^k(n). \tag{6.31}$$

As was the case for the function $h(n)$, we find that the Taylor expansion of $g(n)$ is actually an asymptotic expansion when n is restricted to obey $y_a(n) = o(\sqrt{N})$ as $N \to \infty$. The reason is that s_a is of the order of $\sqrt{N}$, and that the factor $\binom{2k}{k}/2^{2k}$ is bounded. The latter fact follows by using the Stirling formula to derive an asymptotic approximation of this factor as $k \to \infty$. This approximation equals $1/\sqrt{\pi k}$.

We conclude therefore that the first term of the Taylor expansion of $g(n)$ provides an asymptotic approximation when n is restricted as mentioned above. Thus, we get

$$g(n) \sim \frac{1}{\sqrt{\alpha R_0}}, \quad 0 < \alpha \le R_0, \quad y_a(n) = o(\sqrt{N}), \quad N \to \infty. \tag{6.32}$$

The approximations of $h(n)$ and $g(n)$ given in (6.22) and (6.32) can now be inserted into the approximation of ρ_n in (6.12). When we do this, it is important to realize that the two conditions on n for validity of the two approximations, namely that $y_a(n) = o(\sqrt{N})$, are the same. We conclude that the asymptotic approximation of ρ_n given in (6.10) holds. $\qquad \square$

6.2   Three Specific Approximations of ρ_n

This section is used to give three specific approximations of ρ_n. The approximations are valid under different restrictions on n. In these approximations, we use the following notation:

$$\bar{\mu}_1 = \frac{R_0 - 1}{R_0} N, \tag{6.33}$$

$$\bar{\sigma}_1 = \sqrt{\frac{N}{R_0}}, \tag{6.34}$$

$$\beta_1 = \text{sign}(R_0 - 1)\sqrt{2N\left(\log R_0 - \frac{R_0 - 1}{R_0}\right)}, \tag{6.35}$$

$$y_1(n) = \frac{n - \bar{\mu}_1}{\bar{\sigma}_1}, \quad n = 1, 2, \ldots, N, \tag{6.36}$$

$$\bar{\mu}_2 = N \log R_0, \tag{6.37}$$

$$\bar{\sigma}_2 = \sqrt{N}, \tag{6.38}$$

$$\beta_2 = \sqrt{N} \log R_0, \tag{6.39}$$

$$y_2(n) = \frac{n - \bar{\mu}_2}{\bar{\sigma}_2}, \quad n = 1, 2, \ldots, N. \tag{6.40}$$

This notation is also found in Appendix A, where notation that is used throughout the monograph is summarized. As before, $\varphi(y) = \exp(-y^2/2)/\sqrt{2\pi}$ denotes the normal density function.

The results are given by the following theorem:

Theorem 6.2 *The quantity*

$$\rho_n = \frac{1}{R_0} \frac{N!}{(N-n)!} \left(\frac{R_0}{N}\right)^n, \quad n = 1, 2, \ldots, N, \tag{6.41}$$

is approximated as follows:

$$\rho_n \sim \frac{1}{\sqrt{R_0}} \frac{\varphi(y_1(n))}{\varphi(\beta_1)}, \quad y_1(n) = o(\sqrt{N}), \quad 1 \leq n, \quad N \to \infty, \tag{6.42}$$

$$\rho_n \sim \frac{1}{R_0} \frac{\varphi(y_2(n))}{\varphi(\beta_2)}, \quad n = o(N), \quad 1 \leq n, \quad N \to \infty, \tag{6.43}$$

$$\rho_n \sim R_0^{n-1}, \quad n = o(\sqrt{N}), \quad 1 \leq n, \quad N \to \infty. \tag{6.44}$$

Proof. The first two approximations in this theorem follow from the more general result in (6.10), while the third approximation is derived from the second one. The first approximation follows by taking $\alpha = 1$ and the second one by taking $\alpha = R_0$. It is straightforward to see that these two α-values correspond to expansions about the two m_a-values $\bar{\mu}_1$ and zero, respectively.

In the first case, with $\alpha = 1$, we get $m_b = m_a = \bar{\mu}_1$ since $A_1(1) = 1$, and also $s_b = s_a = \sqrt{N/R_0} = \bar{\sigma}_1$. Therefore, we have $y_a(n) = y_b(n) = y_1(n)$. Furthermore, $A_2(1) = 1$, and therefore $h_0 = N[\log R_0 - 1 + 1/R_0] = \beta_1^2/2$. Insertions into the general result in (6.10) show that (6.42) holds.

In the second case, with $\alpha = R_0$, we evaluate first the two functions A_1 and A_2. They are found to take the values $A_1(R_0) = R_0(1 + \log(1/R_0))$ and $A_2(R_0) = R_0(1 + \log(1/R_0) + (\log(1/R_0))^2/2)$. Therefore we get $m_a = 0$, $m_b = (R_0 - A_1(R_0))N/R_0 = N\log R_0 = \bar{\mu}_2$, and $s_a = s_b = \sqrt{N} = \bar{\sigma}_2$. Hence $y_a(n) = (n - m_a)/s_a = n/\sqrt{N}$ and $y_b(n) = (n - m_b)/s_b = (n - \bar{\mu}_2)/\bar{\sigma}_2 = y_2(n)$. Also, $h_0 = N[\log R_0 - 1 + A_2(R_0)/R_0] = \beta_2^2/2$. Insertions of these expressions into the general result in (6.10) shows that the approximation in (6.43) is valid for $n/\sqrt{N} = o(\sqrt{N})$. This condition is clearly equivalent to $n = o(N)$, as claimed in (6.43). Thus (6.43) holds.

The third approximation, (6.44), is a special case of the second one. It holds for a smaller range of n-values. It gives a simple impression about the behavior of ρ_n for small positive values of n. It hinges on the following simple approximation of the ratio $\varphi(y_2(n))/\varphi(\beta_2)$:

$$\frac{\varphi(y_2(n))}{\varphi(\beta_2)} \sim R_0^n, \quad n = o(\sqrt{N}), \quad N \to \infty. \tag{6.45}$$

To derive it we use the identity

$$\frac{\varphi(y_2(n))}{\varphi(\beta_2)} = \exp\left(\frac{\beta_2^2}{2} - \frac{(y_2(n))^2}{2}\right)$$

$$= \exp\left(\frac{1}{2}N(\log R_0)^2 - \frac{1}{2}\left(\frac{n - N\log R_0}{\sqrt{N}}\right)^2\right)$$

$$= \exp\left(n\log R_0 - \frac{n^2}{2N}\right) = R_0^n \exp\left(-\frac{n^2}{2N}\right). \tag{6.46}$$

The last factor of the last expression approaches the value one as $N \to \infty$ if n is of smaller magnitude than $\sqrt{N}$. Thus, the approximation (6.45) holds. $\square$

We proceed to give a description the three results in (6.42)–(6.44).

Note that the first approximation of ρ_n, (6.42), is proportional to a normal density function with the argument $y_1(n)$. The definition $y_1(n) = (n - \bar{\mu}_1)/\bar{\sigma}_1$ shows that the corresponding distribution has its mean equal to the carrying capacity $\bar{\mu}_1$, and its standard deviation equal to $\bar{\sigma}_1$. The condition that $y_1(n) = o(\sqrt{N})$ describes the body and the near-end tails of this normal distribution. The restriction of n to positive integer values has the consequence that the range of n-values for which the first approximation is valid varies strongly with the parameter R_0. Thus, if R_0 has a fixed value larger than one, then $\bar{\mu}_1$ is of the order of N and actually equal to a fixed proportion of N. Since $\bar{\sigma}_1$ is of the order of $\sqrt{N}$ we conclude that the n-values both in the body and in the near-end tails of the distribution all belong to the interval from 1 to N. Thus, the first approximation holds throughout the body and the near-end tails of the normal distribution with mean $\bar{\mu}_1$ and standard deviation $\bar{\sigma}_1$. However, if R_0 is fixed at a value in the interval from 0 to 1, then $\bar{\mu}_1$ is negative, and both the body and the near-end tails of the distribution with mean $\bar{\mu}_1$ and standard deviation $\bar{\sigma}_1$ have negative n-values. But ρ_n is defined only for n-values in the interval $1 \leq n \leq N$. Therefore, the first approximation does not hold for any value of n if $R_0 < 1$. In the transition region, finally, $\bar{\mu}_1 \sim \rho\sqrt{N}$ is of the order of $\sqrt{N}$, and $\bar{\mu}_1/\bar{\sigma}_1 \sim \rho$ is finite as $N \rightarrow \infty$. The body and the near-end tails of the distribution with mean $\bar{\mu}_1$ and standard deviation $\bar{\sigma}_1$ will then contain both positive and negative values of n. The first approximation is then valid for that portion of the body and the near-end tails of this distribution where n is positive.

The second approximation of ρ_n, (6.43) is seen to be proportional to a normal density function with the argument $y_2(n)$. In this case, the corresponding distribution has its mean equal to $\bar{\mu}_2$ and its standard deviation equal to $\bar{\sigma}_2$. The range of n-values for which this approximation is valid lies in the left tail of this normal distribution if $R_0 > 1$, in the right tail if $R_0 < 1$, and in the body if R_0 lies in the transition region.

The third approximation of ρ_n is, as mentioned above, a special case of the second one. It has a simpler form than the second one, at the expense of a shorter interval of validity.

6.3   Approximations of $\sum_{n=1}^{N} \rho_n$

The sum of ρ_n over all n-values from 1 to N behaves in qualitatively different ways in different parts of parameter space. We give explicit asymptotic approximations in the present section. In the description of the results, we use the notation

$$\gamma_1 = \log R_0 - \frac{R_0 - 1}{R_0}, \tag{6.47}$$

which appears also in Appendix A. We note that

$$\gamma_1 \geq 0, \quad R_0 > 0, \tag{6.48}$$

$$\beta_1^2 = 2\gamma_1 N, \tag{6.49}$$

$$\varphi(\beta_1) = \frac{1}{\sqrt{2\pi}} \exp(-\gamma_1 N). \tag{6.50}$$

The results are summarized in the following theorem:

Theorem 6.3 *The quantity $\sum_{n=1}^{N} p_n$, where*

$$p_n = \frac{1}{R_0} \frac{N!}{(N-n)!} \left(\frac{R_0}{N}\right)^n, \quad n = 1, 2, \ldots, N, \tag{6.51}$$

is approximated as follows:

$$\sum_{n=1}^{N} p_n \sim \frac{\sqrt{N}}{R_0 \varphi(\beta_1)} = \frac{\sqrt{2\pi N}}{R_0} \exp(\gamma_1 N), \quad R_0 > 1, \quad N \to \infty, \tag{6.52}$$

$$\sum_{n=1}^{N} p_n \sim \sqrt{N} H_1(\rho), \quad\quad\quad\quad \rho = O(1), \quad N \to \infty, \tag{6.53}$$

$$\sum_{n=1}^{N} p_n \sim \frac{1}{1 - R_0}, \quad\quad\quad\quad R_0 < 1, \quad N \to \infty. \tag{6.54}$$

The derivations of these three approximations will be given in Sects. 6.3.1–6.3.3.

The first of these results shows that the approximation of the sum $\sum_{n=1}^{N} p_n$ grows exponentially with N when $R_0 > 1$. The second result is valid in the transition region near $R_0 = 1$, where ρ defined by $\rho = (R_0 - 1)\sqrt{N}$ remains of the order of one as $N \to \infty$. The arguments for this particular rescaling of R_0 are given later, in Sect. 7.3. The second result shows that the approximation in the transition region is of the order of $\sqrt{N}$ in the transition region. The third approximation, finally, is seen to be of the order of one, independent of N, when $R_0 < 1$. The results in (6.52)–(6.54) are forerunners of several results where qualitatively different behaviors are found in the three parameter regions.

It is possible to summarize all three of these approximations in one formula that is valid uniformly over all three of the parameter regions. This formula reads as follows:

$$\sum_{n=1}^{N} p_n \sim \frac{\sqrt{N}}{R_0} H_1(\beta_{13}), \quad N \to \infty, \tag{6.55}$$

where the function $H_1(y) = \Phi(y)/\varphi(y)$ is given by (5.29), and where β_{13} is defined as follows:

$$\beta_{13} = \begin{cases} \beta_1, & R_0 \geq 1, \\ \beta_3, & R_0 \leq 1. \end{cases} \tag{6.56}$$

Here, β_1 was defined in (6.35), while β_3 is given by

$$\beta_3 = \frac{R_0 - 1}{R_0} \sqrt{N}. \tag{6.57}$$

These definitions are also given in Appendix A. We note that both β_1 and β_3 are equal to zero for $R_0 = 1$, and that both of them have the same sign as $R_0 - 1$. The behaviors of them near $R_0 = 1$ are found by using the reparametrization $R_0 = 1 + \rho/\sqrt{N}$, and studying their asymptotic approximations as $N \to \infty$ while keeping ρ

fixed, which describes the so-called transition region. The results are that both of them are asymptotically equal to ρ. We conclude that

$$\beta_{13} \sim \rho, \quad \rho = O(1), \quad N \to \infty. \tag{6.58}$$

We quote from Corollary 5.3 that $H_1(y)$ has the following asymptotic approximations as $y \to \pm\infty$:

$$H_1(y) \sim \begin{cases} \dfrac{1}{\varphi(y)}, & y \to +\infty, \\[2mm] -\dfrac{1}{y}, & y \to -\infty. \end{cases} \tag{6.59}$$

By using these properties of H_1 and the above expressions and approximations of β_{13}, it is readily seen that the uniform result in (6.55) agrees asymptotically with the separate results given in each of the three parameter regions in (6.52)–(6.54).

The uniform result in (6.55) is of considerable importance, since it is needed in Sect. 8.2.3, where we derive an approximation of $\sum_{n=1}^{N} \pi_n$ in the transition region. We shall see further powerful uses of uniform approximations in Chap. 13.

The uniform result in (6.55) gives a better approximation of $\sum_{n=1}^{N} \rho_n$ than the three separate results in (6.52)–(6.54). In addition, it has the advantage of giving a unique approximation for each value of ρ or R_0, while there are areas of overlap between the three parameter regions that make it unclear which approximation in (6.52)–(6.54) to use in any particular situation. Illustrations are given in Fig. 6.2.

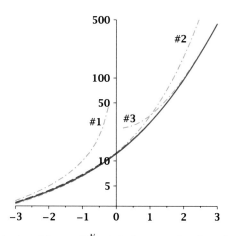

Fig. 6.2 The numerical value of the sum $\sum_{n=1}^{N} \rho_n$ is shown as a function of $\rho = (R_0 - 1)\sqrt{N}$. Comparisons are made with four different explicit approximations, namely the uniform approximation in (6.55), and the three approximations in (6.52)–(6.54). The approximation for $R_0 < 1$ is called #1, the approximation in the transition region is called #2, and the approximation for $R_0 > 1$ is called #3. The uniform approximation is shown *dashed*, while the numerical value is *solid*. The two are practically indistinguishable in the scale used in the figure. The uniform approximation is seen to be superior to the other three approximations. The vertical scale is logarithmic, $N = 100$, and $\mu = 1$

The results in Theorem 6.3 and (6.55) were contained in Nåsell [44]. The derivation given in that paper is different from the derivation that we give in the following three subsections. The former derivation is based on an explicit and exact expression for the sum $\sum_{n=1}^{N} p_n$ in terms of the upper incomplete gamma function $\Gamma(a,x)$ defined by

$$\Gamma(a,x) = \int_x^\infty t^{a-1} \exp(-t)\, dt. \tag{6.60}$$

We refer to Nåsell [44] for details. The method used in the following three subsections has the advantage over the 1996 derivation that it is expected to be more generally applicable, and not restricted to the particular case of the SIS model.

The approximations in this section deal with sums of certain well-defined quantities over n-values from 1 to N. Similar sums will appear in later chapters. Approximations of these sums are derived by partitioning the interval of n-values into contiguous subintervals in which different approximations or bounds are developed. The boundaries for these subintervals are given by the following integers:

$$M_1 = \left[\bar{\sigma}_1^{1/2}\right], \quad M_2 = \left[\bar{\sigma}_1^{3/2}\right], \quad M_3 = \left[\bar{\mu}_1 - \bar{\sigma}_1^{3/2}\right], \quad M_4 = \left[\bar{\mu}_1 + \bar{\sigma}_1^{3/2}\right]. \tag{6.61}$$

Here, as before, brackets around a real number x are used to denote the largest integer smaller than or equal to x. The integers M_3 and M_4 will only be used when $R_0 > 1$. We note that $0 < M_1 < M_2 < M_3 < M_4 < N$ for sufficiently large values of N.

We proceed to prove the three statements in Theorem 6.3 in the following three subsections.

6.3.1 Approximation of $\sum_{n=1}^{N} p_n$ for $R_0 > 1$

We derive here the approximation claimed in (6.52) for $\sum_{n=1}^{N} p_n$ when $R_0 > 1$ as $N \to \infty$. In this case we partition the interval of n-values from 1 to N into three contiguous subintervals, namely $[1, M_3]$, $[M_3 + 1, M_4]$, and $[M_4 + 1, N]$. We note that M_3 lies in the left near-end tail and M_4 lies in the right near-end tail of the distribution with mean $\bar{\mu}_1$ and standard deviation $\bar{\sigma}_1$. To determine this, we use the function z_1, which in line with the usage in Chap. 5 is defined by

$$z_1(n) = \frac{n - \bar{\mu}_1}{\bar{\sigma}_1^2}. \tag{6.62}$$

Evaluating y_1 and z_1 at the two points M_3 and M_4, we get

$$y_1(M_3) = \frac{M_3 - \bar{\mu}_1}{\bar{\sigma}_1} \sim -\bar{\sigma}_1^{1/2}, \tag{6.63}$$

$$z_1(M_3) = \frac{M_3 - \bar{\mu}_1}{\bar{\sigma}_1^2} \sim -\frac{1}{\bar{\sigma}^{1/2}}, \tag{6.64}$$

$$y_1(M_4) = \frac{M_4 - \bar{\mu}_1}{\bar{\sigma}_1} \sim \bar{\sigma}_1^{1/2}, \tag{6.65}$$

$$z_1(M_4) = \frac{M_4 - \bar{\mu}_1}{\bar{\sigma}_1^2} \sim \frac{1}{\bar{\sigma}^{1/2}}. \tag{6.66}$$

Hence, $y_1(M_3) \to -\infty$ and $z_1(M_3) \to 0$ as $\bar{\sigma}_1 \to \infty$, which shows that M_3 lies in the left near-end tail of the distribution with mean $\bar{\mu}_1$ and standard deviation $\bar{\sigma}_1$. Similarly we conclude that M_4 lies in the right near-end tail of the same distribution. The sum $S = \sum_{n=1}^{N} \rho_n$ can be written as a sum of three sums as follows:

$$S = \sum_{n=1}^{N} \rho_n = S_1 + S_2 + S_3, \tag{6.67}$$

where the sums S_1–S_3 are defined by

$$S_1 = \sum_{n=1}^{M_3} \rho_n, \tag{6.68}$$

$$S_2 = \sum_{n=M_3+1}^{M_4} \rho_n, \tag{6.69}$$

$$S_3 = \sum_{n=M_4+1}^{N} \rho_n. \tag{6.70}$$

It is claimed in (6.52) for $R_0 > 1$ that S is approximated by S_0 defined as follows:

$$S_0 = \frac{\sqrt{N}}{R_0 \varphi(\beta_1)}. \tag{6.71}$$

We proceed to prove this claim by establishing the following three results:

- $S_2/S_0 \sim 1$ for $R_0 > 1$ as $N \to \infty$,
- S_1/S_0 is exponentially small for $R_0 > 1$ as $N \to \infty$,
- S_3/S_0 is exponentially small for $R_0 > 1$ as $N \to \infty$.

We note that ρ_n can be approximated by (6.42) for all n-values in the interval $M_3 \leq n \leq M_4$. We use this to derive an approximation of S_2/S_0. By also using the definition of S_0 in (6.71) we find that

$$\frac{S_2}{S_0} \sim \frac{1}{\bar{\sigma}_1} \sum_{n=M_3+1}^{M_4} \varphi(y_1(n)) = \frac{1}{\bar{\sigma}_1} \sum_{n=-\infty}^{M_4} \varphi(y_1(n)) - \frac{1}{\bar{\sigma}_1} \sum_{n=-\infty}^{M_3} \varphi(y_1(n)),$$

$$R_0 > 1, \quad N \to \infty. \tag{6.72}$$

Here we make use of the observation made above that M_4 belongs to the near-end right tail and that M_3 belongs to the near-end left tail of the distribution with mean $\bar{\mu}_1$ and standard deviation $\bar{\sigma}_1$. The two sums in the right-most expression above can therefore be approximated by (5.52). The result is

$$\frac{S_2}{S_0} \sim 1 - \frac{\varphi(y_1(M_4))}{y_1(M_4)} + \frac{\varphi(y_1(M_3))}{y_1(M_3)} \sim 1, \quad R_0 > 1, \quad N \to \infty. \tag{6.73}$$

The conclusion that the ratio S_2/S_0 is asymptotic to one follows from the facts that the argument $y_1(M_4)$ of the first function φ approaches plus infinity and that the argument $y_1(M_3)$ of the second function φ approaches minus infinity as $N \to \infty$, and the fact that normal density functions with arguments of such large absolute values are exponentially small.

It was noted in Chap. 4 that the expression (4.7) can be used to show that the sequence ρ_n is increasing for $1 \le n \le [\bar{\mu}_1]$ and decreasing for $[\bar{\mu}_1] + 1 \le n \le N$ when $R_0 > 1$. We conclude therefore that

$$\rho_n < \rho_{M_3}, \quad n = 1, 2, \dots, M_3 - 1, \tag{6.74}$$

and that

$$\rho_n < \rho_{M_4}, \quad n = M_4 + 1, M_4 + 2, \dots, N. \tag{6.75}$$

Two consequences of these inequalities are that the sums S_1 and S_3 can be bounded as follows:

$$S_1 < M_3 \rho_{M_3} < N \rho_{M_3}, \tag{6.76}$$

and

$$S_3 < (N - M_4) \rho_{M_4} < N \rho_{M_4}. \tag{6.77}$$

The values of ρ_{M_3} and ρ_{M_4} can be approximated by (6.42). It turns out that the approximations of these two quantities are equal, since $y_1(M_3) \sim -\bar{\sigma}_1^{1/2}$ while $y_1(M_4) \sim \bar{\sigma}_1^{1/2}$. We get

$$\rho_{M_i} \sim \frac{1}{\sqrt{R_0}} \frac{\varphi(\bar{\sigma}_1^{1/2})}{\varphi(\beta_1)}, \quad i = 3, 4, \quad N \to \infty. \tag{6.78}$$

By using these approximations of ρ_{M_3} and ρ_{M_4}, we are led to asymptotic approximations of the upper bounds of the sums S_1 and S_3. We denote asymptotic approximations of these sums by $\tilde{S}_1$ and $\tilde{S}_3$, respectively. After divisions by S_0 we get the following bounds:

$$\frac{\tilde{S}_i}{S_0} \le \sqrt{N} \varphi(\bar{\sigma}_1^{1/2}), \quad i = 1, 3, \quad N \to \infty. \tag{6.79}$$

These upper bounds are both exponentially small. This concludes the derivation of the result in (6.52) for $R_0 > 1$.

6.3.2 Approximation of $\sum_{n=1}^{N} \rho_n$ for $R_0 < 1$

We use this subsection to derive the approximation (6.54) in Theorem 6.3.

To derive this approximation, we partition the interval of integer values from one to N into the two subintervals $[1, M_2]$ and $[M_2 + 1, N]$, where $M_2 = \left\lceil \bar{\sigma}_1^{3/2} \right\rceil$ is defined by (6.61). The sum $S = \sum_{n=1}^{N} \rho_n$ can then be written as a sum of two sums S_1 and S_2 defined as follows:

$$S_1 = \sum_{n=1}^{M_2} \rho_n \quad \text{and} \quad S_2 = \sum_{n=M_2+1}^{N} \rho_n, \tag{6.80}$$

with

$$S = S_1 + S_2. \tag{6.81}$$

We proceed to show that $S_1 \sim S_0 = 1/(1 - R_0)$ as $N \to \infty$, and that S_2 is exponentially small.

The n-values from 1 to M_2 are all of smaller magnitude than N. Thus, ρ_n can be approximated by (6.43). We get therefore

$$S_1 = \sum_{n=1}^{M_2} \rho_n \sim \frac{1}{R_0 \varphi(\beta_2)} \sum_{n=1}^{M_2} \varphi(y_2(n)), \quad R_0 < 1, \quad N \to \infty. \tag{6.82}$$

The sum of normal densities can be approximated with the aid of results in Chap. 5. Since $R_0 < 1$, we have $\bar{\mu}_2 < 0$, so both 0 and M_2 are larger than $\bar{\mu}_2$. We show that both 0 and M lie in the far-end right tail of the distribution with mean $\bar{\mu}_2$ and standard deviation $\bar{\sigma}_2$. To show this, we evaluate y_2 and z_2 at these two points, where we define z_2 as follows, in line with previous usage:

$$z_2(n) = \frac{y_2(n)}{\bar{\sigma}_2} = \frac{n - \bar{\mu}_2}{\bar{\sigma}_2^2}, \quad n = 1, 2, \ldots, N. \tag{6.83}$$

We get

$$y_2(M_2) = \frac{M_2 - \bar{\mu}_2}{\bar{\sigma}_2} \sim \frac{(N/R_0)^{3/4} - N \log R_0}{\sqrt{N}} = \sqrt{N} \log \frac{1}{R_0} + \frac{N^{1/4}}{R_0^{3/4}}, \quad N \to \infty, \tag{6.84}$$

$$z_2(M_2) = \frac{y_2(M_2)}{\sqrt{N}} \sim \log \frac{1}{R_0} + \frac{1}{N^{1/4} R_0^{3/4}}, \quad N \to \infty, \tag{6.85}$$

$$y_2(0) = -\frac{\bar{\mu}_2}{\bar{\sigma}_2} = \sqrt{N} \log \frac{1}{R_0} = -\beta_2, \tag{6.86}$$

$$z_2(0) = \frac{y_2(0)}{\sqrt{N}} = \log \frac{1}{R_0}. \tag{6.87}$$

Here, we use the notation $\beta_2 = \sqrt{N} \log R_0$, as in Appendix A. We note that both $y_2(M_2)$ and $y_2(0)$ approach $+\infty$ as $N \to \infty$. Thus, both M_2 and 0 belong to the right tail of the distribution with mean $\bar{\mu}_2$ and standard deviation $\bar{\sigma}_2$. Furthermore, both $z_2(M_2)$ and $z_2(0)$ approach constant non-zero values as $N \to \infty$. We conclude that both M_2 and 0 belong to the right far-end tail of the distribution with mean $\bar{\mu}_2$ and standard deviation $\bar{\sigma}_2$.

This means that (5.53) can be used to approximate the sum of normal densities in the right-hand side of (6.82). We get

$$\sum_{n=-\infty}^{M_2} \varphi(y_2(n)) - \sum_{n=-\infty}^{0} \varphi(y_2(n)) \sim \frac{\varphi(y_2(0))}{\exp(z_2(0)) - 1} - \frac{\varphi(y_2(M_2))}{\exp(z_2(M_2)) - 1},$$
$$R_0 < 1, \quad N \to \infty. \tag{6.88}$$

The second term in this expression is exponentially small compared to the first one. By excluding it, and noting that $y_2(0) = -\beta_2$ and $\exp(z_2(0)) = 1/R_0$, we get

$$S_1 \sim \frac{1}{R_0 \varphi(\beta_2)} \frac{\varphi(\beta_2)}{1/R_0 - 1} = \frac{1}{1 - R_0}, \quad R_0 < 1, \quad N \to \infty, \tag{6.89}$$

as stated above.

We consider now the sum S_2, and show that it is exponentially small. Since $R_0 < 1$, the quantities ρ_n are decreasing for all n. It follows that

$$S_2 = \sum_{n=M_2+1}^{N} \rho_n < \sum_{n=M_2+1}^{N} \rho_{M_2} = (N - M_2)\rho_{M_2} < N\rho_{M_2}, \quad R_0 < 1. \tag{6.90}$$

To show that S_2 is exponentially small, it suffices to show that the upper bound $N\rho_{M_2}$ is exponentially small. By applying (6.43) we get

$$\rho_{M_2} \sim \frac{1}{R_0} \frac{\varphi(y_2(M_2))}{\varphi(\beta_2)}, \quad N \to \infty. \tag{6.91}$$

The upper bound of S_2 can therefore be approximated by

$$N\rho_{M_2} \sim \frac{N}{R_0} \frac{\varphi(y_2(M_2))}{\varphi(\beta_2)} \sim \frac{N}{R_0} \exp\left(-\left(\frac{N}{R_0}\right)^{3/4} \log\left(\frac{1}{R_0}\right) - \frac{1}{2} \frac{N^{1/2}}{R_0^{3/2}}\right) = o(1),$$
$$R_0 < 1, \quad N \to \infty. \tag{6.92}$$

We conclude from this that S_2 is exponentially small for $R_0 < 1$ as $N \to \infty$. This concludes the derivation of the approximation (6.54) of $\sum_{n=1}^{N} \rho_n$ in the parameter region where $R_0 < 1$.

6.3.3   Approximation of $\sum_{n=1}^{N} \rho_n$ in the Transition Region

The so-called transition region near $R_0 = 1$ will be defined in Sect. 7.3. It is given by keeping the parameter $\rho = (R_0 - 1)\sqrt{N}$ constant as $N \to \infty$. In this subsection we derive the approximation (6.53) of $\sum_{n=1}^{N} \rho_n$ in the transition region. The derivation shows similarities with the derivation that was carried out in the previous subsection. Thus, we define $M_2 = \left[\bar{\sigma}_1^{3/2}\right]$ as in (6.61) and use M_2 to partition the interval $[1, N]$ into the two subintervals $[1, M_2]$ and $[M_2 + 1, N]$. The sum $S = \sum_{n=1}^{N} \rho_n$ can then be written

$$S = S_1 + S_2, \tag{6.93}$$

where

$$S_1 = \sum_{n=1}^{M_2} \rho_n \quad \text{and} \quad S_2 = \sum_{n=M_2+1}^{N} \rho_n. \tag{6.94}$$

We show that $S_1 \sim H_1(\rho)\sqrt{N}$, and that S_2 is exponentially small as $\rho = O(1)$ and $N \to \infty$

Now, since $M_2 = o(N)$, we can use the approximation for ρ_n given by (6.43) in S_1. Thus, we get

$$S_1 \sim \frac{1}{R_0 \varphi(\beta_2)} \sum_{n=1}^{M_2} \varphi(y_2(n))$$

$$= \frac{1}{R_0 \varphi(\beta_2)} \left[\sum_{n=-\infty}^{M_2} \varphi(y_2(n)) - \sum_{n=-\infty}^{0} \varphi(y_2(n)) \right],$$

$$\rho = O(1), \quad N \to \infty. \tag{6.95}$$

To approximate the sums of normal densities here, we need to determine the values of $y_2(M_2)$ and $y_2(0)$, and also the corresponding values of z_2 if the point in question lies in a tail of the distribution with mean $\bar{\mu}_2$ and standard deviation $\bar{\sigma}_2$. We get

$$y_2(M_2) = \frac{M_2 - \bar{\mu}_2}{\bar{\sigma}_2} = \frac{M_2}{\sqrt{N}} - \beta_2 \sim \frac{N^{1/4}}{R_0^{3/4}} - \beta_2 \sim N^{1/4} - \rho, \quad \rho = O(1), \quad N \to \infty, \tag{6.96}$$

$$z_2(M_2) = \frac{y_2(M_2)}{\bar{\sigma}_2} \sim N^{-1/4} - \frac{\rho}{\sqrt{N}}, \quad \rho = O(1), \quad N \to \infty, \tag{6.97}$$

$$y_2(0) = \frac{-\bar{\mu}_2}{\bar{\sigma}_2} = -\beta_2 \sim -\rho, \quad \rho = O(1), \quad N \to \infty. \tag{6.98}$$

We conclude from this that $y_2(M_2) \to \infty$ and $z_2(M_2) \to 0$ as $N \to \infty$. This means that the point M_2 belongs to the right near-end tail of the distribution with mean $\bar{\mu}_2$ and standard deviation $\bar{\sigma}_2$. The first sum of normal densities in the last expression of (6.95) is therefore approximated by (5.52). Furthermore, $y_2(0)$ is bounded for large N. Therefore, the point 0 belongs to the body of the distribution with mean $\bar{\mu}_2$ and standard deviation $\bar{\sigma}_2$. The second sum of normal densities above is therefore approximated by (5.51). By applying these approximations we get

$$S_1 \sim \frac{\bar{\sigma}_2}{R_0 \varphi(\beta_2)} \left[1 - \frac{\varphi(y_2(M_2))}{y_2(M_2)} - \left(\Phi(y_2(0)) + \frac{1}{2\bar{\sigma}_2} \varphi(y_2(0)) \right) \right],$$

$$\rho = O(1), \quad N \to \infty. \tag{6.99}$$

This is further simplified by noting that $\beta_2 \sim \rho$ and that $y_2(0) \sim -\rho$. Thus we get

$$S_1 \sim \frac{\sqrt{N}}{R_0 \varphi(\rho)} \left[1 - \frac{\varphi(y_2(M_2))}{y_2(M_2)} - \Phi(-\rho) - \frac{1}{2\sqrt{N}} \varphi(-\rho) \right]$$

$$\sim \frac{\Phi(\rho)}{\varphi(\rho)} \sqrt{N}, \quad \rho = O(1), \quad N \to \infty. \tag{6.100}$$

The last approximation step follows since $\varphi(y_2(M_2))$ is exponentially small, $1 - \Phi(-\rho) = \Phi(\rho)$, and $R_0 \sim 1$. Thus, we have shown that $S_1 \sim H_1(\rho)\sqrt{N}$.

We consider now the sum S_2. We note first that $\bar{\mu}_1 \sim \rho\sqrt{N}$, while $M_2 \sim (N/R_0)^{3/4}$. It follows therefore that $M_2 > \bar{\mu}_1$. We use this inequality to conclude that the quantities ρ_n are decreasing in n for $n \geq M_2$. Thus

$$S_2 = \sum_{n=M_2+1}^{N} \rho_n < \sum_{n=M_2+1}^{N} \rho_{M_2} = (N - M_2)\rho_{M_2} < N\rho_{M_2}. \tag{6.101}$$

The quantity ρ_{M_2} can be approximated by (6.43). We get

$$\rho_{M_2} \sim \frac{1}{R_0} \frac{\varphi(y_2(M_2))}{\varphi(\beta_2)} = \frac{1}{R_0} \frac{\varphi(M_2/\sqrt{N} - \beta_2)}{\varphi(\beta_2)}$$

$$= \frac{1}{R_0} \exp\left(-\frac{1}{2} \left(\frac{M_2}{\sqrt{N}} \right)^2 + \frac{M_2 \beta_2}{\sqrt{N}} \right), \quad \rho = O(1), \quad N \to \infty. \tag{6.102}$$

The exponent in the last expression is asymptotically equal to

$$-\frac{1}{2} \frac{N^{1/2}}{R_0^{3/2}} + \left(\frac{N}{R_0} \right)^{3/4} \log\left(1 + \frac{\rho}{\sqrt{N}} \right) \sim -\frac{1}{2} N^{1/2} + \rho N^{1/4}, \quad \rho = O(1), \quad N \to \infty. \tag{6.103}$$

This shows that the exponent approaches $-\infty$ as $N \to \infty$. We conclude therefore that the sum S_2 is exponentially small. This concludes the derivation of the approximation in (6.53). Hence, all three approximations in Theorem 6.3 have been derived.

As an interesting aside, we note that the function of N defined by $\sum_{n=1}^{N} \rho_n$ in case $R_0 = 1$ is denoted $Q(N)$ by Knuth [38]. Thus,

$$Q(N) = \sum_{n=1}^{N} \frac{N!}{(N-n)!N^n}. \tag{6.104}$$

This function was investigated by Knuth as part of a study of the asymptotic behavior of various algorithms. It is closely related to a problem raised by Ramanujan [63]. For this reason, Flajolet et al. [28] refer to it as "Ramanujan's Q-function". Clearly, we get

$$Q(N) \sim \sqrt{N}H_1(0) = \sqrt{\frac{\pi N}{2}}, \quad N \to \infty, \tag{6.105}$$

as also given by Knuth. Knuth also suggests the investigation of $\sum_{n=1}^{N} \rho_n$ in the two cases $R_0 > 1$ and $R_0 < 1$ in Exercise 1.2.11.3.11 of the quoted reference.

Chapter 7
Approximation of the Stationary Distribution $\mathbf{p}^{(1)}$ of the SIS Model

We use this chapter to derive approximations of the stationary distribution $\mathbf{p}^{(1)}$, while we deal with the stationary distribution $\mathbf{p}^{(0)}$ in Chap. 9. The reason for dealing with $\mathbf{p}^{(1)}$ first is that it is the easy one, and that we actually use some of the results derived for $\mathbf{p}^{(1)}$ in the derivation of approximations for $\mathbf{p}^{(0)}$.

All of the derivations in this chapter are based on the expression (3.8) for the probability $p_n^{(1)}$ in terms of ρ_n, which says that

$$p_n^{(1)} = \frac{\rho_n}{\sum_{n=1}^{N} \rho_n}, \quad n = 1, 2, \ldots, N. \tag{7.1}$$

The preceding chapter was used to derive approximations of both numerator and denominator of this expression. All that is needed here is to insert appropriate approximations from Chap. 6 into the above expression for $p_n^{(1)}$.

7.1 Approximation of the Stationary Distribution $\mathbf{p}^{(1)}$ when R_0 is Distinctly Above One

In this section we give approximations of the stationary distribution $\mathbf{p}^{(1)}$ in the parameter region where R_0 is distinctly above the deterministic threshold value one. Four approximations are given. The first one is valid in the body and the near-end tails of the distribution, while the second and third one are valid in the left tail. The fourth one is not an approximation of the distribution, but rather of the probability $p_1^{(1)}$. The third of the approximations is valid for a smaller range of n-values than the second one, and the fourth one is a simple special case of the third one.

The four approximations are given in the following theorem.

I. Nåsell, *Extinction and Quasi-stationarity in the Stochastic Logistic SIS Model*, Lecture Notes in Mathematics 2022, DOI 10.1007/978-3-642-20530-9_7, © Springer-Verlag Berlin Heidelberg 2011

Theorem 7.1 *The stationary distribution $\mathbf{p}^{(1)}$ of the SIS model is approximated as follows when $R_0 > 1$ as $N \to \infty$:*

$$p_n^{(1)} \sim \frac{1}{\bar{\sigma}_1} \varphi(y_1(n)), \quad R_0 > 1, \quad y_1(n) = o(\sqrt{N}), \quad N \to \infty, \tag{7.2}$$

$$p_n^{(1)} \sim \frac{1}{\sqrt{N}} \varphi(\beta_1) \frac{\varphi(y_2(n))}{\varphi(\beta_2)}, \quad R_0 > 1, \quad n = o(N), \quad 1 \le n, \quad N \to \infty, \tag{7.3}$$

$$p_n^{(1)} \sim \frac{1}{\sqrt{N}} \varphi(\beta_1) R_0^n, \quad R_0 > 1, \quad n = o(\sqrt{N}), \quad 1 \le n, \quad N \to \infty, \tag{7.4}$$

$$p_1^{(1)} \sim \frac{1}{\sqrt{N}} \varphi(\beta_1) R_0, \quad R_0 > 1, \quad N \to \infty. \tag{7.5}$$

Proof. The first three approximations in this theorem are derived by inserting the approximations of ρ_n from (6.42)–(6.44) and the approximation of $\sum_{n=1}^{N} \rho_n$ from (6.52) for $R_0 > 1$ into the (7.1). The fourth approximation is an obvious special case of the third one. $\qquad\square$

The first of these results shows that the distribution $\mathbf{p}^{(1)}$ is approximately normal in its body and its near-end tails. The second result shows that the left tail of the same distribution is approximated by the left tail of another normal distribution, modulo scaling. The third result shows that the left-most tail depends on n in a simple way.

7.2   Approximation of the Stationary Distribution $\mathbf{p}^{(1)}$ when R_0 is Distinctly Below One

All the quantities that we study change drastically when N is large if R_0 is lowered from a fixed value larger than one to a fixed value smaller than one. A main difference concerning the stationary distribution $\mathbf{p}^{(1)}$ is that the probabilities $p_n^{(1)}$ have a maximum as a function of n when $R_0 > 1$, but that they decrease monotonically as a function of n when $R_0 < 1$.

The stationary distribution $\mathbf{p}^{(1)}$ is described by the probabilities $p_n^{(1)}$ for each $n = 1, 2, \ldots, N$. We deal also with the distribution function

$$F^{(1)}(n) = \sum_{k=1}^{n} p_k^{(1)}, \quad n = 1, 2, \ldots, N. \tag{7.6}$$

We give approximations of $p_n^{(1)}$ and $F^{(1)}(n)$ when $R_0 < 1$, N is large, and n is properly restricted. The results are given by the following theorem:

Theorem 7.2 *The stationary distribution $\mathbf{p}^{(1)}$ of the SIS model is approximated as follows when $R_0 < 1$ as $N \to \infty$:*

$$p_n^{(1)} \sim \frac{1-R_0}{R_0} \frac{\varphi(y_2(n))}{\varphi(\beta_2)}, \qquad R_0 < 1, \quad n = o(N), \qquad 1 \le n, \quad N \to \infty, \tag{7.7}$$

$$p_n^{(1)} \sim (1-R_0)R_0^{n-1}, \qquad R_0 < 1, \quad n = o(\sqrt{N}), \quad 1 \le n, \quad N \to \infty, \tag{7.8}$$

$$p_1^{(1)} \sim 1 - R_0, \qquad R_0 < 1, \quad N \to \infty, \tag{7.9}$$

$$F^{(1)}(n) \sim 1 - \frac{\varphi(y_2(n))}{\varphi(\beta_2)}, \qquad R_0 < 1, \quad n = o(N), \qquad 1 \le n, \quad N \to \infty, \tag{7.10}$$

$$F^{(1)}(n) \sim 1 - R_0^n, \qquad R_0 < 1, \quad n = o(\sqrt{N}), \quad 1 \le n, \quad N \to \infty. \tag{7.11}$$

Proof. The first result follows by inserting the approximation (6.43) for ρ_n and the approximation (6.54) for $\sum_{n=1}^{N} \rho_n$ into (7.1). The second result that holds when n is restricted to values that are of smaller order then $\sqrt{N}$ can be derived from the first one by application of the approximation

$$\frac{\varphi(y_2(n))}{\varphi(\beta_2)} \sim R_0^n, \quad n = o(\sqrt{N}), \quad N \to \infty, \tag{7.12}$$

given in (6.45). The third result is an obvious special case of the second one, found by putting $n = 1$. The fourth result needs an argument. The starting point is the approximation of $p_n^{(1)}$ given in (7.7). By applying this approximation, we get

$$F^{(1)}(n) \sim \frac{1-R_0}{R_0} \frac{1}{\varphi(\beta_2)} \sum_{k=1}^{n} \varphi(y_2(k)),$$

$$= \frac{1-R_0}{R_0} \frac{1}{\varphi(\beta_2)} \left[\sum_{k=-\infty}^{n} \varphi(y_2(k)) - \sum_{k=-\infty}^{0} \varphi(y_2(k)) \right],$$

$$R_0 < 1, \quad n = o(N), \quad 1 \le n, \quad N \to \infty. \tag{7.13}$$

The sums of normal densities that appear here are evaluated with the aid of the approximations given in Chap. 5. The evaluations require knowledge about the four quantities $y_2(n)$, $z_2(n)$, $y_2(0)$, and $z_2(0)$. We get

$$y_2(n) = \frac{n - \bar{\mu}_2}{\bar{\sigma}_2} = \frac{n - N\log R_0}{\sqrt{N}} = \frac{n}{\sqrt{N}} + \sqrt{N}\log\frac{1}{R_0}, \tag{7.14}$$

$$z_2(n) = \frac{y_2(n)}{\bar{\sigma}_2} = \frac{n}{N} + \log\frac{1}{R_0}, \tag{7.15}$$

$$y_2(0) = -\frac{\bar{\mu}_2}{\bar{\sigma}_2} = \sqrt{N}\log\frac{1}{R_0} = -\beta_2, \tag{7.16}$$

$$z_2(0) = \frac{y_2(0)}{\bar{\sigma}_2} = \log \frac{1}{R_0}. \tag{7.17}$$

It follows that both $y_2(n)$ and $y_2(0)$ approach $+\infty$ as $N \to \infty$, since $R_0 < 1$. Furthermore, we see that neither $z_2(n)$, nor $z_2(0)$ approach zero as $N \to \infty$. We conclude therefore that both 0 and n belong to the far-end right tail of the distribution with mean $\bar{\mu}_2$ and standard deviation $\bar{\sigma}_2$. Both the sums of normal densities in (7.13) are therefore approximated by the aid of (5.53). Thus, we get

$$\frac{1}{\bar{\sigma}_2} \sum_{k=1}^{n} \varphi(y_2(k)) = \frac{1}{\bar{\sigma}_2} \sum_{k=-\infty}^{n} \varphi(y_2(k)) - \frac{1}{\bar{\sigma}_2} \sum_{k=-\infty}^{0} \varphi(y_2(k))$$

$$\sim 1 - \frac{1}{\bar{\sigma}_2} \frac{\varphi(y_2(n))}{\exp(z_2(n)) - 1} - \left[1 - \frac{1}{\bar{\sigma}_2} \frac{\varphi(y_2(0))}{\exp(z_2(0)) - 1} \right]$$

$$\sim \frac{1}{\bar{\sigma}_2} \left[\frac{\varphi(\beta_2)}{1/R_0 - 1} - \frac{\varphi(y_2(n))}{1/R_0 - 1} \right] = \frac{R_0}{1 - R_0} \frac{\varphi(\beta_2)}{\bar{\sigma}_2} \left[1 - \frac{\varphi(y_2(n))}{\varphi(\beta_2)} \right],$$

$$R_0 < 1, \quad n = o(N), \quad N \to \infty, \tag{7.18}$$

since $\exp(z_2(0)) = 1/R_0$, and $\exp(z_2(n)) \sim 1/R_0$ for $n = o(N)$ as $N \to \infty$. Multiplication by $\bar{\sigma}_2$ and insertion into (7.13) establishes the approximation of the distribution function $F^{(1)}(n)$ given in (7.10). Finally, the fifth and last approximation is derived from the fourth one by application of (6.45).   □

The result in (7.7) shows that the stationary distribution $\mathbf{p}^{(1)}$ is, modulo scaling, approximated by a portion of the right tail of a normal density with negative mean for n-values up to the order of $o(N)$. The left-most part of this approximating distribution is by (7.8) seen to coincide with part of a geometric distribution. The third approximation gives a simple approximation of the probability $p_1^{(1)}$. It is somewhat surprising to see by (7.10) that the distribution function $F^{(1)}$ is approximated by one minus the right tail of a normal density with negative mean, properly scaled. The result in (7.11) is consistent with the finding that the left-most part of the stationary distribution $\mathbf{p}^{(1)}$ is approximately geometric.

7.3 Reparametrization of R_0 in the Transition Region

We recall from (7.2) that the stationary distribution $\mathbf{p}^{(1)}$ is for $R_0 > 1$ and N sufficiently large approximated in its body and its near-end tails by a normal distribution with mean $\bar{\mu}_1$ and standard deviation $\bar{\sigma}_1$. The random variable having this distribution takes only positive values. The normal approximation can therefore be expected to be acceptable only if the ratio of $\bar{\mu}_1$ to $\bar{\sigma}_1$ is large. We find from (6.33) and (6.34) that this ratio equals

$$\frac{\bar{\mu}_1}{\bar{\sigma}_1} = \frac{R_0 - 1}{\sqrt{R_0}} \sqrt{N}. \tag{7.19}$$

The normal distribution is a good approximation if this ratio is large, but poor if the ratio is small. In the asymptotic analysis context we interpret this by saying that the normal distribution is a good approximation if this ratio approaches infinity as $N \to \infty$, and a poor approximation if $\bar{\mu}_1 / \bar{\sigma}_1 = O(1)$ as $N \to \infty$. The expression for the ratio $\bar{\mu}_1 / \bar{\sigma}_1$ above shows that this ratio remains bounded if the quantity

$$\rho = (R_0 - 1)\sqrt{N} \tag{7.20}$$

remains bounded. The introduction of ρ represents a reparametrization of R_0 that makes R_0 a function of N. If ρ remains bounded as $N \to \infty$ then R_0 approaches one. The parameter region described by the condition $\rho = O(1)$ defines the transition region that will be studied repeatedly for all the quantities that we are concerned with for the stochastic logistic SIS model. We note the rule of thumb that the boundary between the transition region and the region where R_0 is distinctly larger than one is for practical purposes approximately given by $\rho = 3$. This rule is motivated by the fact that a normally distributed random variable takes values that deviate by more than three standard deviations from its mean with small probability.

The reparametrization of R_0 in the transition region has also been derived by Dolgoarshinnykh et al. [25], using a diffusion process approximation.

To describe the results in the transition region, we introduce the following notation:

$$\bar{\mu}_3 = \rho\sqrt{N}, \tag{7.21}$$

$$\bar{\sigma}_3 = \sqrt{N}, \tag{7.22}$$

$$y_3(n) = \frac{n - \bar{\mu}_3}{\bar{\sigma}_3}. \tag{7.23}$$

This notation is also found in Appendix A.

It is useful to note that the several quantities introduced in (6.33)–(6.40), and (6.57) obey the following simple asymptotic relations in the transition region:

$$\bar{\mu}_1 \sim \bar{\mu}_3, \quad \bar{\sigma}_1 \sim \bar{\sigma}_3, \quad \beta_1 \sim \rho, \quad \bar{\mu}_2 \sim \bar{\mu}_3, \quad \bar{\sigma}_2 = \bar{\sigma}_3, \quad \beta_2 \sim \rho, \quad \beta_3 \sim \rho,$$
$$\rho = O(1), \quad N \to \infty. \tag{7.24}$$

Also, the functions y_1 and y_2 are asymptotically equivalent to the function y_3 in the sense that

$$y_1(n) \sim y_3(n), \quad y_2(n) \sim y_3(n), \quad n = 1, 2, \ldots, N, \quad \rho = O(1), \quad N \to \infty. \tag{7.25}$$

7.4   Approximation of the Stationary Distribution $\mathbf{p}^{(1)}$ in the Transition Region

We give two approximations of the stationary distribution $p_n^{(1)}$ and two correspond-ing approximations of the distribution function $F^{(1)}(n) = \sum_{k=1}^n p_k^{(1)}$ in the transition region. The results are summarized in the following theorem.

Theorem 7.3 *The stationary distribution* $\mathbf{p}^{(1)}$ *of the SIS model is approximated as follows in the transition region:*

$$p_n^{(1)} \sim \frac{1}{\Phi(\rho)} \frac{1}{\bar{\sigma}_3} \varphi(y_3(n)), \quad \rho = O(1), \quad n = o(N), \quad 1 \le n, \quad N \to \infty, \quad (7.26)$$

$$p_n^{(1)} \sim \frac{1}{H_1(\rho)\sqrt{N}}, \quad \rho = O(1), \quad n = o(\sqrt{N}), \quad 1 \le n, \quad N \to \infty, \quad (7.27)$$

$$F^{(1)}(n) \sim \frac{1}{\Phi(\rho)} \left[\Phi(y_3(n)) - \Phi(y_3(0)) \right],$$

$$\rho = O(1), \quad n = o(N), \quad 1 \le n, \quad N \to \infty, \quad (7.28)$$

$$F^{(1)}(n) \sim \frac{n}{H_1(\rho)\sqrt{N}}, \quad \rho = O(1), \quad n = o(\sqrt{N}), \quad 1 \le n, \quad N \to \infty. \quad (7.29)$$

Proof. It is illuminating to note that the result in (7.26) can be derived in two different ways. The reason for this is that the restriction to the parameter region that we refer to as the transition region implies that the left tail described by $n = o(N)$ coincides with the body and the near-end tails where $y_1(n) = o(\sqrt{N})$. This implies that we can use either (6.42) or (6.43) to derive an approximation of ρ_n. Thus, using (6.42), we get

$$\rho_n \sim \frac{1}{\sqrt{R_0}} \frac{\varphi(y_1(n))}{\varphi(\beta_1)} \sim \frac{\varphi(y_3(n))}{\varphi(\rho)}, \quad \rho = O(1), \quad n = o(N), \quad 1 \le n, \quad N \to \infty, \quad (7.30)$$

using the approximations $\beta_1 \sim \rho$ and $y_1(n) \sim y_3(n)$ in (7.24) and (7.25), and the fact that $R_0 \sim 1$ when $\rho = O(1)$. If on the other hand we use (6.43), we get

$$\rho_n \sim \frac{1}{R_0} \frac{\varphi(y_2(n))}{\varphi(\beta_2)} \sim \frac{\varphi(y_3(n))}{\varphi(\rho)}, \quad \rho = O(1), \quad n = o(N), \quad 1 \le n, \quad N \to \infty, \quad (7.31)$$

using the approximations of $\beta_2 \sim \rho$ and $y_2(n) \sim y_3(n)$ in (7.24) and (7.25), and also the approximation $R_0 \sim 1$, which is valid in the transition region. Thus, we get the same approximation of ρ_n either way.

The result in (7.26) follows by inserting this approximation of ρ_n and the approximation of $\sum_{n=1}^N \rho_n$ for $\rho = O(1)$ in (6.53) into (7.1).

The result in (7.27) follows from (7.26) by noting that

$$\frac{\varphi(y_3(n))}{\varphi(\rho)} = \exp\left(\frac{1}{2}\rho^2 - \frac{1}{2}\left(\rho - \frac{n}{\sqrt{N}}\right)^2\right)$$

$$= \exp\left(\frac{\rho n}{\sqrt{N}} - \frac{1}{2}\frac{n^2}{N}\right) \sim 1, \quad \rho = O(1), \quad n = o(\sqrt{N}), \quad N \to \infty, \quad (7.32)$$

and recalling that $H_1(y) = \Phi(y)/\varphi(y)$.

The corresponding approximations of the distribution function $F^{(1)}(n)$ are derived from the approximation of $p_n^{(1)}$ in (7.26). We get

$$F^{(1)}(n) = \sum_{k=1}^{n} p_k^{(1)} \sim \frac{1}{\Phi(\rho)}\frac{1}{\bar{\sigma}_3}\sum_{k=1}^{n}\varphi(y_3(k))$$

$$= \frac{1}{\Phi(\rho)}\left[\frac{1}{\bar{\sigma}_3}\sum_{k=-\infty}^{n}\varphi(y_3(k)) - \frac{1}{\bar{\sigma}_3}\sum_{k=-\infty}^{0}\varphi(y_3(k))\right],$$

$$\rho = O(1), \quad n = o(N), \quad 1 \leq n, \quad N \to \infty. \quad (7.33)$$

The two sums here are both approximated by the aid of (5.49). We define

$$z_3(n) = \frac{y_3(n)}{\bar{\sigma}_3}, \quad (7.34)$$

and note that $y_3(n) = n/\sqrt{N} - \rho$ and that $z_3(n) = n/N - \rho/\sqrt{N}$. For any $n = o(N)$ we get $z_3(n) = o(1)$, which gives $G_1(z_3(n)) \sim 1/2$. This includes also the case when $n = 0$. Applying (5.49) twice, we get

$$F^{(1)}(n) \sim \frac{1}{\Phi(\rho)}\left[\Phi(y_3(n)) + \frac{1}{2\bar{\sigma}_3}\varphi(y_3(n)) - \Phi(y_3(0)) - \frac{1}{2\bar{\sigma}_3}\varphi(y_3(0))\right]$$

$$\sim \frac{1}{\Phi(\rho)}\left[\Phi(y_3(n)) - \Phi(y_3(0))\right],$$

$$\rho = O(1), \quad 1 \leq n, \quad n = o(N), \quad N \to \infty. \quad (7.35)$$

This shows that (7.28) holds. The approximation of the distribution function in the left-most tail given in (7.29) follows from (7.28) by application of the approximation (5.27). An alternate derivation is of course to insert the constant approximation of $p_n^{(1)}$ in (7.27) into the sum $\sum_{k=1}^{n} p_k^{(1)}$ that defines $F^{(1)}(n)$. □

It is noteworthy that the approximation of $p_n^{(1)}$ in its left-most tail given in (7.27) is independent of n. This does not mean that $p_n^{(1)}$ itself is independent of n, only that an additional term of smaller magnitude in its asymptotic approximation is required in order to detect the n-dependence.

Chapter 8
Preparations for the Study of the Stationary Distribution $\mathbf{p}^{(0)}$ of the SIS Model

This chapter prepares for the approximations of the stationary distribution $\mathbf{p}^{(0)}$ of the SIS model that are derived in Chap. 9. We recall from (3.7) that the stationary distribution $p_n^{(0)}$ is determined from the sequence π_n by

$$p_n^{(0)} = \frac{\pi_n}{\sum_{n=1}^{N} \pi_n}, \quad n = 1, 2, \ldots, N. \tag{8.1}$$

We give approximations of the numerator π_n in Sect. 8.1, and approximations of the denominator $\sum_{n=1}^{N} \pi_n$ in Sect. 8.2. Our derivations rely heavily on the results in Chap. 6, where approximations of ρ_n and of $\sum_{n=1}^{N} \rho_n$ are derived. As in that chapter, we note that the approximations of $\sum_{n=1}^{N} \pi_n$ are far more intricate than the approximations of π_n.

8.1 Approximation of π_n

In this section we establish three asymptotic approximations of π_n, as follows:

Theorem 8.1 *The quantity*

$$\pi_n = \frac{1}{nR_0} \frac{N!}{(N-n)!} \left(\frac{R_0}{N}\right)^n, \quad n = 1, 2, \ldots, N, \tag{8.2}$$

is approximated as follows:

$$\pi_n \sim \frac{1}{(R_0 - 1)\sqrt{N}\varphi(\beta_1)} \frac{1}{\bar{\sigma}_1} \varphi(y_1(n)),$$

$$R_0 > 1, \quad y_1(n) = o(\sqrt{N}), \quad 1 \leq n, \quad N \to \infty, \tag{8.3}$$

I. Nåsell, *Extinction and Quasi-stationarity in the Stochastic Logistic SIS Model*, Lecture Notes in Mathematics 2022, DOI 10.1007/978-3-642-20530-9_8, © Springer-Verlag Berlin Heidelberg 2011

$$\pi_n \sim \frac{1}{nR_0}\frac{\varphi(y_2(n))}{\varphi(\beta_2)}, \quad n=o(N), \quad 1 \le n, \quad N \to \infty, \qquad (8.4)$$

$$\pi_n \sim \frac{1}{n}R_0^{n-1}, \quad n=o(\sqrt{N}), \quad 1 \le n, \quad N \to \infty. \qquad (8.5)$$

Proof. These three approximations follow from the corresponding approximations of ρ_n in (6.42)–(6.44) by using the relation between $\pi_n = \rho_n/n$ given in (4.9). For the first one we note also that the identity

$$n = \bar{\mu}_1 + (n - \bar{\mu}_1) \qquad (8.6)$$

leads to the asymptotic approximation

$$n \sim \bar{\mu}_1, \quad \frac{n - \bar{\mu}_1}{\bar{\mu}_1} = o(1), \quad N \to \infty, \qquad (8.7)$$

and also that the condition for validity of this approximation is equivalent to the condition $y_1(n) = o(\sqrt{N})$ as $N \to \infty$. □

In addition, we establish a relation between the two sequences π_n and ρ_n that makes use of their dependence on R_0. We emphasize this functional dependence by writing $\pi_n(R_0)$ and $\rho_n(R_0)$ instead of π_n and ρ_n. The relation says that the value of $\pi_n(R_0)$ can be expressed with the aid of an integral of $\rho_n(R_0)$ over R_0. The specific relation is as follows:

$$\pi_n(R_0) = \frac{1}{R_0}\int_0^{R_0}\rho_n(x)\,dx, \quad n=1,2,\ldots,N. \qquad (8.8)$$

This relation is basic in the rather touchy derivation of an approximation of $\sum_{n=1}^{N}\pi_n$ in the transition region in Sect. 8.2.3.

The derivation of (8.8) is based on the explicit expression (4.6) for ρ_n. This expression has the important feature that the dependence on R_0 is simple. It can be written $\rho_n = \rho_n(R_0) = CR_0^{n-1}$, where C depends on N and n, but not on R_0. By multiplying with R_0 and using the relation $\pi_n = \rho_n/n$ in (4.9), we find that $R_0\pi_n(R_0) = CR_0^n/n$. It follows by differentiation that

$$\frac{d}{dR_0}(R_0\pi_n(R_0)) = CR_0^{n-1} = \rho_n(R_0). \qquad (8.9)$$

The relation (8.8) follows since $R_0\pi_n(R_0)$ takes the value 0 for $R_0 = 0$.

8.2   Approximations of $\sum_{n=1}^{N} \pi_n$

The approximation of the sum $\sum_{n=1}^{N} \pi_n$ in the transition region contains a function H_0 that is defined here. Its definition depends on a negative constant ρ_b, which is specified first. The recommendation is to take ρ_b in the interval from -6 to -3. Comments on the choice of ρ_b are given below. In the numerical evaluations we use $\rho_b = -3$. This is also the choice in the Maple module described in Appendix B. With ρ_b given, H_0 is defined by

$$H_0(y) = \begin{cases} H_a(y), & y \le \rho_b, \\ H_a(\rho_b) + \displaystyle\int_{\rho_b}^{y} H_1(t)\,dt, & \rho > \rho_b, \end{cases} \tag{8.10}$$

where the auxiliary function H_a is defined by

$$H_a(y) = -\log|y| + \sum_{k=1}^{m_d} (-1)^k \frac{a_k}{2k} \frac{1}{y^{2k}}, \quad y < 0, \tag{8.11}$$

and m_d is a positive integer depending on ρ_b. Its values for the indicated range of ρ_b-values are given in Table 8.1 in Sect. 8.2.3. In particular, we note that $m_d = 4$ for $\rho_b = -3$. These are the values that are used in numerical illustrations, and also in the Maple procedures in Appendix B. The coefficients a_k are defined by (5.35), where they are used to describe the asymptotic approximation of $H_1(y)$ as $y \to -\infty$. We recall that the first few relevant ones are $a_1 = 1$, $a_2 = 3$, $a_3 = 15$, $a_4 = 105$. We note that H_0 is plotted together with H_1 in Fig. 8.1, and also in Fig. 13.1. Both functions grow slowly for negative arguments and quickly for positive arguments. It is also seen that $H_0(y)$ is smaller than $H_1(y)$.

Like the sum $\sum_{n=1}^{N} p_n$, we find that the sum $\sum_{n=1}^{N} \pi_n$ behaves in qualitatively different ways in the three parameter regions. We derive the three results given in the following theorem:

Theorem 8.2 *The quantity* $\sum_{n=1}^{N} \pi_n$, *where*

$$\pi_n = \frac{1}{nR_0} \frac{N!}{(N-n)!} \left(\frac{R_0}{N}\right)^n, \quad n = 1, 2, \ldots, N, \tag{8.12}$$

is approximated as follows:

$$\sum_{n=1}^{N} \pi_n \sim \frac{1}{(R_0 - 1)\sqrt{N}\varphi(\beta_1)}, \quad R_0 > 1, \quad N \to \infty, \tag{8.13}$$

$$\sum_{n=1}^{N} \pi_n \approx \frac{1}{2}\log N + H_0(\rho), \quad \rho = O(1), \quad N \to \infty, \tag{8.14}$$

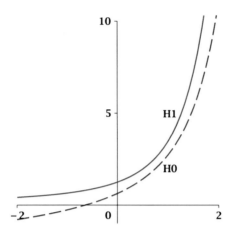

Fig. 8.1 The functions $H_1(y)$ and $H_0(y)$ are shown as functions of y

$$\sum_{n=1}^{N} \pi_n \sim \frac{1}{R_0} \log \frac{1}{1 - R_0}, \qquad R_0 < 1, \qquad N \to \infty. \qquad (8.15)$$

The derivations of the three approximations given in this theorem will occupy us in the rest of this chapter. The approximations given for $R_0 > 1$ and for $R_0 < 1$ are easy to derive, and we give them first, in Sects. 8.2.1 and 8.2.2, respectively. The approximation in the transition region is more demanding, and will be given last, in Sect. 8.2.3.

The results given in (8.13)–(8.15) show that the approximation of the sum $\sum_{n=1}^{N} \pi_n$ grows exponentially with N when $R_0 > 1$, that it is of the order of $\log N$ in the transition region, and that it is independent of N for $R_0 < 1$. One can compare this with the corresponding results for $\sum_{n=1}^{N} \rho_n$ in (6.52)–(6.54). It is clear from the relation $\pi_n = \rho_n/n$ in (4.9) that $\pi_n < \rho_n$ for $n > 1$, and that therefore also $\sum_{n=1}^{N} \pi_n < \sum_{n=1}^{N} \rho_n$. The ratio between them, $\sum_{n=1}^{N} \pi_n / \sum_{n=1}^{N} \rho_n$, is found to be of the order of $1/N$ when $R_0 > 1$, of the order of $\log N / \sqrt{N}$ in the transition region, and of the order of one when $R_0 < 1$.

8.2.1   Approximation of $\sum_{n=1}^{N} \pi_n$ for $R_0 > 1$

We show in this subsection that (8.13) holds. The derivation makes use of the results in Sect. 6.3.1, where an asymptotic approximation of $\sum_{n=1}^{N} \rho_n$ is derived for $R_0 > 1$ as $N \to \infty$. We use notation similar to that in Sect. 6.3.1. In particular, we partition the interval from 1 to N into three contiguous subintervals with the aid of the two integers M_3 and M_4 defined in (6.61). We use sums that resemble the sums S, S_1, etc. that are used in Sect. 6.3.1. The corresponding sums that we use here involve π_n instead of ρ_n, and are denoted by primes. They are defined as follows:

$$S' = \sum_{n=1}^{N} \pi_n, \tag{8.16}$$

$$S_1' = \sum_{n=1}^{M_3} \pi_n. \tag{8.17}$$

$$S_2' = \sum_{n=M_3+1}^{M_4} \pi_n, \tag{8.18}$$

$$S_3' = \sum_{n=M_4+1}^{N} \pi_n, \tag{8.19}$$

$$S_0' = \frac{1}{(R_0 - 1)\sqrt{N}\varphi(\beta_1)}. \tag{8.20}$$

We proceed to establish the following three results:

- $S_2'/S_0' \sim 1$ for $R_0 > 1$ as $N \to \infty$,
- S_1'/S_0' is exponentially small for $R_0 > 1$ as $N \to \infty$,
- S_3'/S_0' is exponentially small for $R_0 > 1$ as $N \to \infty$.

The derivation makes use of the results for the corresponding unprimed sums in Sect. 6.3.1.

To prove that $S_2'/S_0' \sim 1$, we note from Theorems 6.2 and 8.1 that $\rho_n/\pi_n \sim \bar{\mu}_1$, $M_3 \leq n \leq M_4$. Hence

$$\frac{S_2}{S_2'} \sim \bar{\mu}_1, \quad R_0 > 1, \quad N \to \infty. \tag{8.21}$$

Since also $S_0/S_0' = \bar{\mu}_1$, we get

$$\frac{S_2'}{S_0'} \sim \frac{S_2}{S_0}, \quad R_0 > 1, \quad N \to \infty, \tag{8.22}$$

and since this ratio was shown in Sect. 6.3.1 to be asymptotic to one, the assertion follows.

To prove that S_1'/S_0' is exponentially small, we note that

$$S_1' = \sum_{n=1}^{M_3} \pi_n = \sum_{n=1}^{M_3} \frac{1}{n}\rho_n < \sum_{n=1}^{M_3} \rho_n = S_1. \tag{8.23}$$

It follows that

$$\frac{S_1'}{S_0'} < \frac{S_1}{S_0'} = \bar{\mu}_1 \frac{S_1}{S_0}. \tag{8.24}$$

The assertion follows, since it was shown in Sect. 6.3.1 that S_1/S_0 is exponentially small, and since $\bar{\mu}_1$ is of the order of N.

To prove that S_3'/S_0' is exponentially small, a similar argument can be used. Thus,

$$S_3' = \sum_{n=M_4+1}^{N} \pi_n = \sum_{n=M_4+1}^{N} \frac{1}{n} \rho_n < \frac{1}{M_4} \sum_{n=M_4+1}^{N} \rho_n = \frac{1}{M_4} S_3. \tag{8.25}$$

It follows that

$$\frac{S_3'}{S_0'} < \frac{S_3}{M_4 S_0'} = \frac{\bar{\mu}_1 S_3}{M_4 S_0} < \frac{S_3}{S_0}, \tag{8.26}$$

using the fact that $\bar{\mu}_1 < M_4$. The assertion follows from this, since it was shown in Sect. 6.3.1 that S_3/S_0 is exponentially small.

Thus, the approximation (8.13) has been derived.

8.2.2 Approximation of $\sum_{n=1}^{N} \pi_n$ for $R_0 < 1$

We show in this subsection that the approximation (8.15) holds. To show this, we write for brevity $\pi_{1,n} = R_0^{n-1}/n$. It follows then that $\pi_n < \pi_{1,n}$ for $n = 2, 3, \ldots, N$, since

$$\frac{\pi_{n+1}}{\pi_n} = \frac{\lambda_n}{\mu_{n+1}} = \frac{n}{n+1} R_0 \left(1 - \frac{n}{N} \right), \tag{8.27}$$

while

$$\frac{\pi_{1,n+1}}{\pi_{1,n}} = \frac{n}{n+1} R_0 > \frac{\pi_{n+1}}{\pi_n}, \tag{8.28}$$

and $\pi_1 = \pi_{1,1} = 1$.

The sum $\sum_{n=1}^{N} \pi_n$ can be written as a sum of two sums as follows:

$$\sum_{n=1}^{N} \pi_n = S_1 + S_2, \tag{8.29}$$

where

$$S_1 = \sum_{n=1}^{M_1} \pi_n \quad \text{and} \quad S_2 = \sum_{n=M_1+1}^{N} \pi_n, \tag{8.30}$$

and where we use $M_1 = \left[\bar{\sigma}_1^{1/2} \right]$ defined in (6.61). We evaluate S_1 using the approximation $\pi_n \sim \pi_{1,n}$ in (8.5), and determine a bound for S_2 using the inequality $\pi_n < \pi_{1,n}$. To evaluate S_1 we note from (8.5) that $\pi_n \sim \pi_{1,n}$ for $1 \leq n \leq M_1$. Hence

$$S_1 \sim \sum_{n=1}^{M_1} \frac{1}{n} R_0^{n-1} = \frac{1}{R_0} \log \frac{1}{1-R_0} - \left[L(R_0, 1, M_1) - \frac{1}{M_1} \right] R_0^{M_1-1}, \tag{8.31}$$

where L denotes the LerchPhi function, defined by

$$L(R_0, a, M) = \sum_{n=0}^{\infty} \frac{R_0^n}{(M+n)^a}. \tag{8.32}$$

Furthermore, S_2 is bounded as follows:

$$S_2 < \sum_{n=M_1+1}^{N} \frac{1}{n} R_0^{n-1} < \sum_{n=M_1+1}^{\infty} \frac{1}{n} R_0^{n-1} = \left[L(R_0, 1, M_1) - \frac{1}{M_1} \right] R_0^{M_1-1}. \tag{8.33}$$

For positive integer values of M we find that

$$L(R_0, 1, M) = \sum_{n=0}^{\infty} \frac{R_0^n}{M+n} < \sum_{n=0}^{\infty} \frac{R_0^n}{M} = \frac{1}{M} \frac{1}{1-R_0}. \tag{8.34}$$

A consequence of this is that the upper bound of S_2 and the correction term for S_1 are both bounded by $R_0^{M_1} / (M_1(1-R_0))$. This bound approaches zero as $N \to \infty$ since then $M_1 \to \infty$ and $R_0 < 1$.

We conclude therefore that the asymptotic approximation of $\sum_{n=1}^{N} \pi_n$ claimed in (8.15) holds.

8.2.3   Approximation of $\sum_{n=1}^{N} \pi_n$ in the Transition Region

We derive the approximation (8.14) in this subsection. This approximation is not asymptotic, and we shall therefore be concerned with the error involved in using this approximation. The derivation of this result is more demanding than the derivations in the two preceding sections. One reason for this is that the "natural" approach of using the approximation (8.4) of π_n, which holds for n-values up to an order smaller than N, will not succeed in giving a useful approximation of $\sum_{n=1}^{N} \pi_n$ in the transition region. This approach would lead us to evaluate a sum of the form

$$\sum_{n=1}^{M_2} \pi_n \sim \frac{1}{R_0 \varphi(\beta_2)} \sum_{n=1}^{M_2} \frac{1}{n} \varphi(y_2(n)), \quad \rho = O(1), \quad N \to \infty, \tag{8.35}$$

with M_2 defined in (6.61). We have not been able to find a useful approximation of this sum.

The failure of this approach causes us to use an alternative method, based on the relation between $\pi_n(R_0)$ and $\rho_n(R_0)$ given in (8.8), which reads

$$\pi_n(R_0) = \frac{1}{R_0} \int_0^{R_0} \rho_n(x)\, dx. \tag{8.36}$$

We note first that addition over all n-values from 1 to N leads to the following expression for $\sum_{n=1}^{N} \pi_n$:

$$\sum_{n=1}^{N} \pi_n = \frac{1}{R_0} \int_0^{R_0} \sum_{n=1}^{N} \rho_n(x)\,dx. \qquad (8.37)$$

The integration here is carried out over the interval of x-values from zero to R_0. This means that we integrate over two parameter regions, namely both the one where R_0 is distinctly below one, and the transition region. Thus, in order to proceed, we need an approximation of the integrand that is valid over both of these parameter regions. The uniform result that is given in (6.55) satisfies this need. It can be simplified slightly in the present case to take the form

$$\sum_{n=1}^{N} \rho_n \sim \frac{\sqrt{N}}{R_0} H_1(\beta_3), \quad R_0 < 1 \quad \text{or} \quad \rho = O(1), \quad N \to \infty. \qquad (8.38)$$

The reason for this simplification is that the quantity β_{13} that appears in (6.55) is by its definition in (6.56) equal to β_1 when $R_0 \geq 1$, and equal to β_3 for $R_0 \leq 1$. In the case that we are dealing with, any R_0-values larger than one will fall in the transition region, where by (7.24) both β_1 and β_3 are asymptotically equal to ρ, and therefore $\beta_1 \sim \beta_3$. By using the above approximation of the integrand, and noting the definition of β_3 in (6.57), we get the following approximation of $\sum_{n=1}^{N} \pi_n$ in the transition region:

$$\sum_{n=1}^{N} \pi_n \sim \frac{\sqrt{N}}{R_0} \int_0^{R_0} \frac{1}{x} H_1\left(\frac{x-1}{x}\sqrt{N}\right) dx, \quad \rho = O(1), \quad N \to \infty. \qquad (8.39)$$

A change of integration variable gives

$$\sum_{n=1}^{N} \pi_n \sim \frac{1}{R_0} \int_{-\infty}^{\beta_3} \frac{1}{1 - y/\sqrt{N}} H_1(y)\,dy, \quad \rho = O(1), \quad N \to \infty. \qquad (8.40)$$

This approximation can be simplified to the following form:

$$\sum_{n=1}^{N} \pi_n \sim \int_{-\infty}^{\rho} \frac{1}{1 - y/\sqrt{N}} H_1(y)\,dy, \quad \rho = O(1), \quad N \to \infty, \qquad (8.41)$$

since $R_0 \sim 1$ and $\beta_3 \sim \rho$ in the transition region, and since the integrand is $O(1)$ at the right-hand end of the integration interval.

We note that the integrand is a product of two factors that both are increasing functions of y. The first factor, $1/(1 - y/\sqrt{N})$, approaches zero as $y \to -\infty$, while it is asymptotically equal to one at the right-hand end of the integration interval. The second factor, $H_1(y)$, is asymptotically equal to $-1/y$ for $y \to -\infty$, as shown by

Corollary 5.3, while it is a well-known function of the order of one at the right-hand end of the integration interval.

We proceed to determine an approximation of the integral in (8.41). To do this, we choose ρ_b to be a negative constant, independent of N, and use it to partition the integration interval $(-\infty, \rho)$ into the union of two subintervals $(-\infty, \rho_b)$ and (ρ_b, ρ) (assuming that $\rho_b < \rho$). Different approximations of the integrand are used in these two subintervals. In the first one, the function $H_1(y)$ is replaced by an asymptotic approximation as $y \to -\infty$, while the factor $1/(1 - y/\sqrt{N})$ is kept as it is, after which the integration is carried out. (This procedure is used also if $\rho < \rho_b$.) In the second interval, roles are reversed: The factor $1/(1 - y/\sqrt{N})$ is replaced by its asymptotic approximation one, while the factor $H_1(y)$ is unchanged.

The approximation of the integrand in the first subinterval will be asymptotic in part of the interval, but it will fail to be asymptotic at the right-hand end, since y is finite there. The resulting approximation of the corresponding integral will therefore also fail to be asymptotic. The error introduced in this way can however be kept acceptably small, as shown below.

Here, we use properties of the function H_1 derived in Chap. 5, Sect. 5.3. In particular, we recall from (5.42) that the sum of the first $m + 1$ terms of the asymptotic approximation of the function $H_1(y) = \Phi(y)/\varphi(y)$ as $y \to -\infty$ can be written

$$A_m(y) = \sum_{k=0}^{m} (-1)^{k+1} \frac{a_k}{y^{2k+1}} = -\frac{1}{y} + \frac{1}{y^3} - \frac{3}{y^5} + \cdots + \frac{(-1)^{m+1} a_m}{y^{2m+1}}, \quad m = 0, 1, 2, \ldots,$$
(8.42)

with a_k given in (5.35). Furthermore, we recall that the expression $A_m(y)$ gives an upper bound (lower bound) of $H_1(y)$ for $y < 0$ if m is even (odd). By applying the bounds we get

$$|H_1(y) - A_m(y)| < -\frac{a_{m+1}}{y^{2m+3}}, \quad m = 0, 1, 2, \ldots, \quad y < 0.$$
(8.43)

With this background we turn to the derivation of the approximation (8.14). Assume first that $\rho > \rho_b$. The integral in (8.41) can then be written as a sum of three integrals as follows:

$$\int_{-\infty}^{\rho} \frac{1}{1 - y/\sqrt{N}} H_1(y) \, dy = I_1 + I_2 + I_3,$$
(8.44)

where

$$I_1 = \int_{-\infty}^{\rho_b} \frac{1}{1 - y/\sqrt{N}} A_m(y) \, dy,$$
(8.45)

$$I_2 = \int_{-\infty}^{\rho_b} \frac{1}{1 - y/\sqrt{N}} (H_1(y) - A_m(y)) \, dy,$$
(8.46)

$$I_3 = \int_{\rho_b}^{\rho} \frac{1}{1 - y/\sqrt{N}} H_1(y)\,dy. \tag{8.47}$$

Here, m is a nonnegative integer that depends on ρ_b and that will be determined later.

Among the three integrals above, the first one can be evaluated explicitly since the integrand is rational, the absolute value of the second one can be bounded with the help of the inequality (8.43), and the integrand of the third one can be approximated by $H_1(y)$, since the factor $1/(1 - y/\bar{\sigma}_2)$ is asymptotically equal to 1.

By inserting the expression (8.42) for $A_m(y)$ into the definition for the integral I_1, we get

$$I_1 = \sum_{k=0}^{m} (-1)^{k+1} a_k J_k, \quad m = 0, 1, 2, \ldots, \tag{8.48}$$

where J_k denotes the integral defined as follows:

$$J_k = \int_{-\infty}^{\rho_b} f_k(y)\,dy, \quad k = 0, 1, 2, \ldots, \tag{8.49}$$

and where the integrand in the integral defining J_k equals

$$f_k(y) = \frac{1}{1 - y/\sqrt{N}} \frac{1}{y^{2k+1}}, \quad k = 0, 1, 2, \ldots. \tag{8.50}$$

We note also that an upper bound for $|I_2|$ can be expressed by using the inequality (8.43). We get

$$|I_2| < -a_{m+1} \int_{-\infty}^{\rho_b} \frac{1}{1 - y/\sqrt{N}} \frac{1}{y^{2m+3}}\,dy = -a_{m+1} J_{m+1}, \quad m = 0, 1, 2, \ldots. \tag{8.51}$$

To proceed, we evaluate the integrals J_k. Actually, we shall be satisfied with asymptotic approximations of these integrals as $N \to \infty$. We begin with partial fraction expansions of the integrands $f_k(y)$. We get

$$f_0(y) = \frac{1}{1 - y/\sqrt{N}} \frac{1}{y} = \frac{1}{y} + \frac{1}{1 - y/\sqrt{N}} \frac{1}{\sqrt{N}}, \tag{8.52}$$

$$f_1(y) = \frac{1}{1 - y/\sqrt{N}} \frac{1}{y^3} = \frac{1}{y^3} + \frac{1}{y^2} \frac{1}{\sqrt{N}} + f_1(y) \frac{1}{N} \sim \frac{1}{y^3}, \quad N \to \infty, \tag{8.53}$$

and

$$f_k(y) \sim \frac{1}{y^{2k+1}}, \quad k = 1, 2, 3, \ldots, \quad N \to \infty. \tag{8.54}$$

To determine J_0 we note first that the indefinite integral of $f_0(y)$ can be evaluated as follows, using the partial fraction expansion given above:

$$\int f_0(y)\,dy = -\log\left(\frac{1}{\sqrt{N}} - \frac{1}{y}\right), \quad y < 0. \tag{8.55}$$

This expression can be used to derive an asymptotic approximation of J_0 as follows:

$$J_0 = \int_{-\infty}^{\rho_b} f_0(y)\,dy = -\log\left(\frac{1}{\sqrt{N}} - \frac{1}{\rho_b}\right) + \log\left(\frac{1}{\sqrt{N}}\right) = -\log\left(\frac{\sqrt{N}}{|\rho_b|} + 1\right)$$

$$= -\log\frac{\sqrt{N}}{|\rho_b|} - \log\left(1 + \frac{|\rho_b|}{\sqrt{N}}\right) \sim -\log\frac{\sqrt{N}}{|\rho_b|} - \frac{|\rho_b|}{\sqrt{N}} \sim -\log\frac{\sqrt{N}}{|\rho_b|}, \quad N \to \infty. \tag{8.56}$$

Asymptotic approximations of the integrals J_k for $k \geq 1$ are easy to determine. We get

$$J_k \sim \int_{-\infty}^{\rho_b} \frac{1}{y^{2k+1}}\,dy = -\frac{1}{2k}\frac{1}{\rho_b^{2k}}, \quad k = 1,2,3,\dots, \quad N \to \infty. \tag{8.57}$$

By using these asymptotic approximations of the integrals J_k, we are led to the following asymptotic approximation of the integral I_1 after putting $a_0 = 1$:

$$I_1 \sim \log\frac{\sqrt{N}}{|\rho_b|} + \sum_{k=1}^{m} \frac{(-1)^k a_k}{2k}\frac{1}{\rho_b^{2k}}$$

$$= \frac{1}{2}\log N - \log|\rho_b| + \sum_{k=1}^{m} \frac{(-1)^k a_k}{2k}\frac{1}{\rho_b^{2k}},$$

$$m = 1,2,\dots, \quad N \to \infty. \tag{8.58}$$

Furthermore, we introduce d_m to denote the asymptotic approximation of the upper bound of $|I_2|$. Thus, we have

$$|I_2| < -a_{m+1} J_{m+1} \sim d_m, \quad m = 0,1,2,\dots, \quad N \to \infty, \tag{8.59}$$

where

$$d_m = \frac{a_{m+1}}{(2m+2)\rho_b^{2m+2}} = \frac{(2m+1)a_m}{(2m+2)\rho_b^{2m+2}}, \quad m = 0,1,2,\dots. \tag{8.60}$$

This approximation of the upper bound of $|I_2|$ depends on ρ_b and on m. It is somewhat surprising that it has a minimum as a function of m for fixed ρ_b. By choosing m equal to this minimum value for fixed ρ_b, the error introduced by ignoring I_2 is as small is possible. To determine the m-value that minimizes d_m, we solve the equation $d_{m+1} = d_m$ for m with fixed ρ_b. This equation is easily established. By using the definition of a_k in (5.35) we get

$$\frac{d_{m+1}}{d_m} = \frac{(2m+3)(m+1)}{(m+2)\rho_b^2}. \tag{8.61}$$

The equation $d_{m+1} = d_m$ leads to a second degree equation in m. Its solution can be written

$$m = \frac{\rho_b^2 - 5}{4} \pm \frac{1}{4}\sqrt{\rho_b^4 + 6\rho_b^2 + 1}. \tag{8.62}$$

There are two roots here, one positive and the other one negative. They are in general not integers. The actual m-value that minimizes d_m for fixed ρ_b is chosen as the integer part of the positive root above, or as the integer part of the positive root, increased by one. The choice between these two possibilities is made by numerical evaluations of the two cases, followed by choosing the one case that leads to the smallest value of d_m. The fact that we do not have a unique formula for determining the minimizing value of m is not very disturbing, since there are only a few cases to be considered.

The results of evaluations for four different values of ρ_b are given in Table 8.1.

The table shows that the asymptotic approximation of the bound for I_2 can be made acceptably small by choosing the absolute value of ρ_b sufficiently large. As mentioned previously, we use $\rho_b = -3$ and $m_d = 4$ in the numerical evaluations, and in the Maple module listed in Appendix B.

The integral I_3, finally, is found to have the following asymptotic approximation:

$$I_3 \sim \int_{\rho_b}^{\rho} H_1(y)\,dy, \quad \rho = O(1), \quad N \to \infty. \tag{8.63}$$

We can now summarize our results as follows:

$$\sum_{n=1}^{N} \pi_n = I_1 + I_2 + I_3 \sim \frac{1}{2}\log N + H_0(\rho) + I_2$$

$$\approx \frac{1}{2}\log N + H_0(\rho), \quad \rho = O(1), \quad N \to \infty. \tag{8.64}$$

The second of these expressions for $\sum_{n=1}^{N} \pi_n$ is a more precise way of writing the approximation (8.14).

Table 8.1 The first column lists four values of ρ_b, for which evaluations have been made. The second column shows the positive root m_+ of the equation (8.62), and the third column shows the value m_d that minimizes the asymptotic approximation d_m of the upper bound of the absolute value of the integral I_2. The asymptotic approximation of the error bound for $|I_2|$ is shown in the last column

ρ_b	m_+	m_d	d_{m_d}
-3	3.9	4	2×10^{-3}
-4	7.4	8	3×10^{-5}
-5	12.0	12	2×10^{-7}
-6	17.5	18	6×10^{-10}

The development so far is based on the assumption that $\rho_b < \rho$. The treatment of the case $\rho \leq \rho_b$ is closely similar. In this case the integral in the left-hand side of (8.44) is equal to the sum of two integrals, denoted I_1' and I_2'. These integrals are found from I_1 and I_2, respectively, by replacing the upper bounds of integration ρ_b by ρ. The absolute value of I_2' is, as above, asymptotically bounded by d_{m_d}, while the integral I_1' is asymptotically equal to

$$\frac{1}{2} \log N + H_a(\rho),\tag{8.65}$$

consistent with the definition of H_0 in (8.10).

This completes the derivation of the approximation (8.14) of $\sum_{n=1}^{N} \pi_n$ in the transition region.

Chapter 9
Approximation of the Stationary Distribution $\mathbf{p}^{(0)}$ of the SIS Model

The stationary distribution $\mathbf{p}^{(0)}$ is by (3.7) determined from the π_n by the formula

$$p_n^{(0)} = \frac{\pi_n}{\sum_{n=1}^{N} \pi_n}, \quad n = 1, 2, \ldots, N. \tag{9.1}$$

Our approximations of the stationary distribution $\mathbf{p}^{(0)}$ are all based on this relation and the approximations of the numerator π_n given in (8.3)–(8.5), together with the approximations of the denominator $\sum_{n=1}^{N} \pi_n$ given in (8.13)–(8.15).

We give separate approximations in each of the three parameter regions.

9.1 Approximation of the Stationary Distribution $\mathbf{p}^{(0)}$ when R_0 is Distinctly Above One

We give three approximations of the stationary distribution $\mathbf{p}^{(0)}$ when R_0 is distinctly above one, one in the body and near-end tails of the distribution, and two in the left tail. In addition, we give an approximation of the important probability $p_1^{(0)}$.

Theorem 9.1 *The stationary distribution $\mathbf{p}^{(0)}$ of the SIS model is approximated as follows when $R_0 > 1$ as $N \to \infty$:*

$$p_n^{(0)} \sim \frac{1}{\bar{\sigma}_1} \varphi(y_1(n)), \qquad R_0 > 1, \quad y_1(n) = o(\sqrt{N}), \quad 1 \leq n, \quad N \to \infty, \tag{9.2}$$

$$p_n^{(0)} \sim \beta_3 \frac{\varphi(\beta_1)}{\varphi(\beta_2)} \frac{\varphi(y_2(n))}{n}, \qquad R_0 > 1, \quad n = o(N), \qquad\qquad 1 \leq n, \quad N \to \infty, \tag{9.3}$$

$$p_n^{(0)} \sim \beta_3 \varphi(\beta_1) \frac{R_0^n}{n}, \qquad R_0 > 1, \quad n = o(\sqrt{N}), \qquad 1 \leq n, \quad N \to \infty, \tag{9.4}$$

I. Nåsell, *Extinction and Quasi-stationarity in the Stochastic Logistic SIS Model*, 115
Lecture Notes in Mathematics 2022, DOI 10.1007/978-3-642-20530-9_9,
© Springer-Verlag Berlin Heidelberg 2011

$$p_1^{(0)} \sim (R_0 - 1)\varphi(\beta_1)\sqrt{N}, \quad R_0 > 1, \quad N \to \infty. \tag{9.5}$$

Proof. The first three of these results follow directly from the approximations of π_n in (8.3)–(8.5) and the approximation of $\sum_{n=1}^{N} \pi_n$ for $R_0 > 1$ in (8.13). For the second and third of the approximations, we use also the definition of β_3 in (6.57), given also in Appendix A. The fourth approximation is a simple special case of the third one, found by setting $n = 1$.   □

A comparison of the first of these approximations with (7.2) shows that the bodies and the near-end tails of the two distributions $\mathbf{p}^{(1)}$ and $\mathbf{p}^{(0)}$ are approximated by the same normal distribution. The expectations of the corresponding random variables are both approximately equal to $\bar{\mu}_1$. We recall from (4.19) that these expectations obey the inequality $EI^{(0)} < EI^{(1)}$. The apparent contradiction is easily resolved by the fact that additional terms in the asymptotic approximations are needed to discover the difference between the two expectations.

The ratio between $p_n^{(0)}$ and $p_n^{(1)}$ is by (9.3) and (7.3) found to be approximated as follows:

$$\frac{p_n^{(0)}}{p_n^{(1)}} \sim \frac{\bar{\mu}_1}{n}, \quad R_0 > 1, \quad n = o(N), \quad 1 \leq n, \quad N \to \infty. \tag{9.6}$$

This is consistent with the exact expression for this ratio expressed by (4.14), which says that

$$\frac{p_n^{(0)}}{p_n^{(1)}} = \frac{1}{n} EI^{(0)}. \tag{9.7}$$

The latter expression is seen to be valid without restrictions on R_0 or N or n.

The left-most tails of the distributions $\mathbf{p}^{(0)}$ and $\mathbf{p}^{(1)}$ can be seen to behave quite differently when $1 < R_0 < 2$. By using (7.4) and (9.4) we get

$$\frac{p_{n+1}^{(1)}}{p_n^{(1)}} \sim R_0, \qquad R_0 > 1, \quad n = o(\sqrt{N}), \quad 1 \leq n, \quad N \to \infty, \tag{9.8}$$

$$\frac{p_{n+1}^{(0)}}{p_n^{(0)}} \sim \frac{n}{n+1} R_0, \quad R_0 > 1, \quad n = o(\sqrt{N}), \quad 1 \leq n, \quad N \to \infty. \tag{9.9}$$

This shows that the probabilities $p_n^{(1)}$ will (asymptotically) increase monotonically with n in the left-most tail of the distribution whenever $R_0 > 1$. In contrast to this, we find that the probabilities $p_n^{(0)}$ will (asymptotically) decrease with n for small n-values if $1 < R_0 < 2$. This feature is illustrated in Fig. 4.1 for the R_0-values 1.5 and 1.3 and 1.1. This behavior is consistent with the n-dependence of ρ_n and π_n noted in Chap. 4.

9.2   Approximation of the Stationary Distribution p$^{(0)}$ when R_0 is Distinctly Below One

We give two approximations of the stationary distribution $\mathbf{p}^{(0)}$ when R_0 is distinctly below one, and in addition an approximation of the important probability $p_1^{(0)}$.

Theorem 9.2 *The stationary distribution* $\mathbf{p}^{(0)}$ *of the SIS model is approximated as follows when* $R_0 < 1$ *as* $N \to \infty$:

$$p_n^{(0)} \sim \frac{1}{\log(1/(1-R_0))} \frac{1}{n} \frac{\varphi(y_2(n))}{\varphi(\beta_2)}, \quad R_0 < 1, \quad n = o(N), \quad 1 \leq n, \quad N \to \infty, \tag{9.10}$$

$$p_n^{(0)} \sim \frac{R_0}{\log(1/(1-R_0))} \frac{1}{n} R_0^{n-1}, \quad R_0 < 1, \quad n = o(\sqrt{N}), \quad 1 \leq n, \quad N \to \infty, \tag{9.11}$$

$$p_1^{(0)} \sim \frac{R_0}{\log(1/(1-R_0))}, \quad R_0 < 1, \quad N \to \infty. \tag{9.12}$$

Proof. The first two of these approximations are found from the approximations of π_n in (8.4) and (8.5), and the approximation of $\sum_{n=1}^{N} \pi_n$ for $R_0 < 1$ in (8.15). The third approximation is an obvious special case of the second one. $\square$

We note that the second of these approximations shows that the left-most tail of the stationary distribution $\mathbf{p}^{(0)}$ is in this case approximated by the so-called log series distribution.

9.3   Approximation of the Stationary Distribution p$^{(0)}$ in the Transition Region

We give two approximations of the stationary distribution $\mathbf{p}^{(0)}$ in the transition region, and in addition an approximation of the important probability $p_1^{(0)}$.

Theorem 9.3 *The stationary distribution* $\mathbf{p}^{(0)}$ *is approximated as follows in the transition region:*

$$p_n^{(0)} \approx \frac{1}{\frac{1}{2}\log N + H_0(\rho)} \frac{1}{\varphi(\rho)} \frac{\varphi(y_3(n))}{n},$$

$$\rho = O(1), \quad n = o(N), \quad 1 \leq n, \quad N \to \infty, \tag{9.13}$$

$$p_n^{(0)} \approx \frac{1}{\frac{1}{2}\log(N) + H_0(\rho)} \frac{1}{n}, \quad \rho = O(1), \quad n = o(\sqrt{N}), \quad 1 \leq n, \quad N \to \infty, \quad (9.14)$$

$$p_1^{(0)} \approx \frac{1}{\frac{1}{2}\log N + H_0(\rho)}, \quad \rho = O(1), \quad N \to \infty. \qquad (9.15)$$

Proof. The first of these two approximations is found by using the approximation of π_n in (8.4) and the approximation of $\sum_{n=1}^{N} \pi_n$ in (8.14). In addition, we use $R_0 \sim 1$, $\beta_2 \sim \rho$, and $y_2(n) \sim y_3(n)$.

The second of the two approximations uses (8.5) instead of (8.4) to approximate π_n. In addition, we note that $R_0^n \sim 1$ for $n = o(\sqrt{N})$:

$$R_0^n = \exp(n \log R_0) = \exp\left(n \log\left(1 + \frac{\rho}{\sqrt{N}}\right)\right). \qquad (9.16)$$

The exponent here is approximated by

$$n \log\left(1 + \frac{\rho}{\sqrt{N}}\right) \sim \frac{n\rho}{\sqrt{N}}, \quad N \to \infty. \qquad (9.17)$$

This means that the exponent approaches zero as $N \to \infty$ if n is restricted to positive values that are of smaller order than $\sqrt{N}$. This proves the claim that $R_0^n \sim 1$ for $n = o(\sqrt{N})$.

The third approximation is a simple special case of the second one, found by putting $n = 1$. ☐

Note that the expression $\frac{1}{2}\log N + H_0(\rho)$ is asymptotically approximated by its first term for N sufficiently large. But the slow growth of the first term with N makes this one-term approximation useless for reasonable values of N, and necessitates the inclusion of both terms.

Chapter 10
Approximation of Some Images Under Ψ for the SIS Model

It was mentioned in Chap. 3, Sect. 3.6, that Ferrari et al. [27] have proved an important result concerning the map Ψ, namely that if ν is an arbitrary discrete distribution with the state space $\{1, 2, \ldots, N\}$, then the sequence ν, $\Psi(\nu)$, $\Psi^2(\nu)$, $\ldots$, converges to the quasi-stationary distribution $\mathbf{q}$. Thus, application of the map Ψ to any vector ν can be expected to move it closer to $\mathbf{q}$. In this chapter we shall study several images under the map Ψ. In particular, we shall apply Ψ to the two distributions $\mathbf{p}^{(1)}$ and $\mathbf{p}^{(0)}$ for which approximations have been derived in Chaps. 7 and 9. The resulting distributions $\Psi\left(\mathbf{p}^{(1)}\right)$ and $\Psi\left(\mathbf{p}^{(0)}\right)$ will both be studied in Sect. 10.1, where we are concerned with the parameter region where R_0 is distinctly above one. In Sect. 10.2 we proceed to the parameter region where R_0 is distinctly below one. In that case we shall only study the distribution $\Psi\left(\mathbf{p}^{(1)}\right)$. Sections 10.3 and 10.4 are both concerned with the transition region where $\rho = O(1)$. In the first of these two sections we shall again study the distribution $\Psi\left(\mathbf{p}^{(1)}\right)$. Finally, the sequence of distributions $\Psi^i\left(\mathbf{p}^{(1)}\right)$, with $i = 1, 2, \ldots$, will be studied in Sect. 10.4, where i is the number of times that the map Ψ has been applied to $\mathbf{p}^{(1)}$.

The results that we establish in this chapter will be applied in Chap. 11, where we deal with approximations of the most important distribution in the monograph, namely the quasi-stationary distribution $\mathbf{q}$ of the SIS model.

The images under the map Ψ of the two stationary distributions $\mathbf{p}^{(1)}$ and $\mathbf{p}^{(0)}$ are denoted $\tilde{\mathbf{p}}^{(1)}$ and $\tilde{\mathbf{p}}^{(0)}$, respectively, as in Sect. 3.9. Thus, we write

$$\tilde{\mathbf{p}}^{(l)} = \Psi\left(\mathbf{p}^{(l)}\right), \quad l = 0, 1. \tag{10.1}$$

The components of the two probability vectors $\tilde{\mathbf{p}}^{(1)}$ and $\tilde{\mathbf{p}}^{(0)}$ are spelled out as follows, using the expressions in (3.109) and (3.110):

$$\tilde{p}_n^{(l)} = \pi_n S_n^{(l)} \tilde{p}_1^{(l)}, \quad l = 0, 1, \quad n = 1, 2, \ldots, N, \tag{10.2}$$

I. Nåsell, *Extinction and Quasi-stationarity in the Stochastic Logistic SIS Model*, Lecture Notes in Mathematics 2022, DOI 10.1007/978-3-642-20530-9_10, © Springer-Verlag Berlin Heidelberg 2011

and where the factors $S_n^{(1)}$ and $S_n^{(0)}$ are defined as sums of the quantities $r_k^{(1)}$ and $r_k^{(0)}$, as follows:

$$S_n^{(l)} = \sum_{k=1}^{n} r_k^{(l)}, \quad l = 0, 1, \quad n = 1, 2, \ldots, N, \tag{10.3}$$

with the ratios $r_k^{(1)}$ and $r_k^{(0)}$ given by

$$r_k^{(l)} = \frac{1 - \sum_{j=1}^{k-1} p_j^{(l)}}{\rho_k}, \quad l = 0, 1, \quad k = 1, 2, \ldots, N. \tag{10.4}$$

We note also that the probabilities $\tilde{p}_1^{(1)}$ and $\tilde{p}_1^{(0)}$ are determined from the following relations:

$$\tilde{p}_1^{(l)} = \frac{1}{\sum_{n=1}^{N} \pi_n S_n^{(l)}}, \quad l = 0, 1. \tag{10.5}$$

These expressions are derived from the requirements that the probabilities $\tilde{p}_n^{(1)}$ and $\tilde{p}_n^{(0)}$ add to one.

We recall from Sect. 3.9 that the two ratios $r_k^{(1)}$ and $r_k^{(0)}$ are decreasing functions of k for the SIS model. We shall use these properties of $r_k^{(1)}$ and $r_k^{(0)}$ in the present chapter.

10.1 Approximations of $\Psi\left(\mathbf{p}^{(1)}\right)$ and $\Psi\left(\mathbf{p}^{(0)}\right)$ when R_0 is Distinctly Above One

The two probability vectors $\Psi\left(\mathbf{p}^{(1)}\right)$ and $\Psi\left(\mathbf{p}^{(0)}\right)$ are clearly different. However, we derive here asymptotic approximations that coincide in the parameter region where R_0 is distinctly above one. As in (6.61), we use $M_1 = \left[\bar{\sigma}_1^{1/2}\right]$. The results are summarized in the following theorem:

Theorem 10.1 *The distributions $\tilde{\mathbf{p}}^{(1)} = \Psi\left(\mathbf{p}^{(1)}\right)$ and $\tilde{\mathbf{p}}^{(0)} = \Psi\left(\mathbf{p}^{(0)}\right)$ for the SIS model are approximated as follows when $R_0 > 1$ as $N \to \infty$:*

$$\tilde{p}_n^{(l)} \sim \frac{1}{\bar{\sigma}_1} \varphi(y_1(n)), \quad l = 0, 1, \quad R_0 > 1, \quad y_1(n) = o(\sqrt{N}), \quad N \to \infty, \tag{10.6}$$

$$\tilde{p}_n^{(l)} \sim \frac{R_0 - 1}{R_0} \varphi(\beta_1) \sqrt{N} \frac{1}{n} \left(\frac{\varphi(y_2(n))}{\varphi(\beta_2)} - 1 \right),$$

$$l = 0, 1, \quad R_0 > 1, \quad n \geq 1, \quad n = o(N), \quad N \to \infty, \tag{10.7}$$

$$\tilde{p}_n^{(l)} \sim \frac{R_0-1}{R_0}\varphi(\beta_1)\sqrt{N}\frac{1}{n}\frac{\varphi(y_2(n))}{\varphi(\beta_2)},$$

$$l=0,1, \quad R_0>1, \quad n>M_1, \quad n=o(N), \quad N\to\infty, \quad (10.8)$$

$$\tilde{p}_n^{(l)} \sim \frac{R_0-1}{R_0}\varphi(\beta_1)\sqrt{N}\frac{R_0^n-1}{n},$$

$$l=0,1, \quad R_0>1, \quad n\geq 1, \quad n=o(\sqrt{N}), \quad N\to\infty, \quad (10.9)$$

$$\tilde{p}_1^{(l)} \sim \frac{(R_0-1)^2}{R_0}\varphi(\beta_1)\sqrt{N}, \quad l=0,1, \quad R_0>1, \quad N\to\infty. \quad (10.10)$$

The derivation of these results will occupy the remainder of Sect. 10.1.

To prove the first result, we use the stochastic ordering result in (3.106). Since by the result in Sect. 3.5, likelihood ratio ordering implies stochastic ordering, we conclude that

$$\mathbf{p}^{(0)} \prec_{ST} \tilde{\mathbf{p}}^{(0)} \prec_{ST} \mathbf{q} \prec_{ST} \tilde{\mathbf{p}}^{(1)} \prec_{ST} \mathbf{p}^{(1)}. \quad (10.11)$$

This says that the two distributions $\tilde{\mathbf{p}}^{(0)}$ and $\tilde{\mathbf{p}}^{(1)}$ are stochastically bounded between $\mathbf{p}^{(0)}$ and $\mathbf{p}^{(1)}$. For the latter two distributions we find from (7.2) and (9.2) that the bodies and the near-end tails have the same asymptotic approximations as $N\to\infty$ in the case considered here, namely $R_0>1$. This approximation therefore holds also for the bodies and the near-end tails of the two distributions $\tilde{\mathbf{p}}^{(l)}, l=0,1$, as claimed in (10.6).

The second approximation deals with the left tails of the two distributions $\tilde{\mathbf{p}}^{(l)}$, $l=0,1$, when $R_0>1$. This approximation can be regarded as a large deviation result. Special cases are given in (10.8) and (10.9). The first of these special cases uses the fact that the ratio $\varphi(y_2(n))/\varphi(\beta_2)$ is an increasing function of n for n in the interval $0<n<\bar{\mu}_2$. Thus, this ratio is larger than the value it takes when $n=M_1$ if n lies in the interval $M_1<n<\bar{\mu}_2$. By using (6.45), we find that $\varphi(y_2(M_1))/\varphi(\beta_2) \sim R_0^{M_1}$. This value approaches infinity as $N\to\infty$ for $R_0>1$. Thus, $\varphi(y_2(n))/\varphi(\beta_2)-1 \sim \varphi(y_2(n))/\varphi(\beta_2)$ under the conditions listed in (10.8), which implies that (10.8) holds. The approximation (10.9) in the left-most tail is derived from the more generally valid approximation of the wider left tail in (10.7) by use of the asymptotic approximation of the ratio $\varphi(y_2(n))/\varphi(\beta_2) \sim R_0^n$ in (6.45). Thus, (10.9) holds. The approximations of the probabilities $\tilde{p}_1^{(l)}, l=0,1$ in (10.10) follow from the approximations of the left-most tails in (10.9) by putting $n=1$. We conclude that the three approximations in (10.8), (10.9), and (10.10) all follow from (10.7). It remains therefore to derive the approximations in (10.7). However, it turns out that the approximations of the probabilities $\tilde{p}_1^{(l)}, l=0,1$, in (10.10) are required in the derivation of the approximations of the left tails of the distributions $\tilde{\mathbf{p}}^{(l)}, l=0,1$, in (10.7). We shall therefore derive the approximations in (10.10) first, and leave the derivation of the approximation in (10.7) till the end.

We proceed to this derivation. Four steps are required. The first step is concerned with finding approximations of the ratios $r_k^{(l)}$ defined in (10.4), while the second one gives approximations of the sums $S_n^{(l)}$ defined in (10.3), and the third step uses (10.5) to give the approximations of $\tilde{p}_1^{(l)}$ that are given in (10.10). It is clear that after these three steps, approximations have been found of the second and third of the three factors that determine $\tilde{p}_n^{(l)}$ in (10.2). Since an approximation of the first factor π_n is available in (8.4), we can use (10.2) in the fourth step to derive the approximations given in (10.7) of the left tails of the distributions $\tilde{\mathbf{p}}^{(l)}$.

We use the two integers M_1 and M_3 defined in (6.61). Thus, we have

$$M_1 = \left[\bar{\sigma}_1^{1/2}\right], \quad M_3 = \left[\bar{\mu}_1 - \bar{\sigma}_1^{3/2}\right]. \tag{10.12}$$

The first step in the derivation leads to the following approximation and bound for $r_k^{(l)}$ both for $l = 0$ and for $l = 1$:

$$r_k^{(l)} \sim \frac{1}{\rho_k}, \quad l = 0, 1, \quad 1 \le k \le M_3, \quad R_0 > 1, \quad N \to \infty, \tag{10.13}$$

$$r_k^{(l)} \le r_{M_1}^{(l)}, \quad l = 0, 1, \quad M_1 \le k \le N, \quad R_0 > 1, \quad N \to \infty, \tag{10.14}$$

where the upper bounds $r_{M_1}^{(l)}$ have the following asymptotic approximation:

$$r_{M_1}^{(l)} \sim \frac{1}{R_0^{M_1-1}}, \quad l = 0, 1, \quad N \to \infty. \tag{10.15}$$

Note that there is a large overlap between those k-values where the approximation in (10.13) is valid, and those for which the upper bound in (10.14) holds.

To prove the validity of the approximation (10.13) for $l = 1$, we note from (10.4) that the numerator of the ratio $r_k^{(1)}$ equals one minus the sum $\sum_{j=1}^{k-1} p_j^{(1)}$. The probabilities $p_j^{(1)}$ are increasing in j for $j \le M_3$ since ρ_j is. Hence, by taking $k \le M_3$, they are all smaller than $p_{M_3}^{(1)}$. Since M_3 lies in the near-end tail of the distribution with mean $\bar{\mu}_1$ and standard deviation $\bar{\sigma}_1$, the probability $p_{M_3}^{(1)}$ is approximated by $\varphi(y_1(M_3))/\bar{\sigma}_1$, using (7.2). Now using the definition of M_3, we find that $y_1(M_3) \sim -\bar{\sigma}_1^{1/2}$. Hence, the upper bound of the probabilities $p_j^{(1)}$ is $\varphi(y_1(M_3))/\bar{\sigma}_1 \sim \exp(-\bar{\sigma}_1/2)/(\sqrt{2\pi}\bar{\sigma}_1)$. The asymptotic approximation of the upper bound is thus exponentially small as $N \to \infty$. It follows that the sum $\sum_{j=1}^{k-1} p_j^{(1)}$ is also exponentially small for $k \le M_3$. We conclude in particular that the numerator of $r_k^{(1)}$ is asymptotically equal to one for $k \le M_3$. This establishes the validity of the approximation of $r_k^{(1)}$ in (10.13).

It is straightforward to use similar arguments to show that the same approximation holds also for $l = 0$.

The inequalities in (10.14) for $l = 0$ and $l = 1$ follow from the facts that $r_k^{(l)}$ decrease in k for both l-values, established in Sect. 3.9.

The approximations of the upper bounds $r_{M_1}^{(l)}$ given in (10.15) follow by first applying (10.13) to conclude that $r_{M_1}^{(l)} \sim 1/\rho_{M_1}$, and then applying the approximation $\rho_k \sim R_0^{k-1}$ given in (6.44). This approximation is valid since $M_1 = \mathrm{o}(\sqrt{N})$.

The second step in the derivation is to establish that $S_n^{(l)}$ can be approximated as follows both for $l = 0$ and for $l = 1$:

$$S_n^{(l)} \sim \frac{R_0}{R_0 - 1}\left(1 - \frac{\varphi(\beta_2)}{\varphi(y_2(n))}\right),$$

$$l = 0, 1, \quad 1 \le n, \quad n = \mathrm{o}(N), \quad R_0 > 1, \quad N \to \infty, \quad (10.16)$$

$$S_n^{(l)} \sim \frac{R_0}{R_0 - 1}\left(1 - \frac{1}{R_0^n}\right), \quad l = 0, 1, \quad 1 \le n \le M_1, \quad R_0 > 1, \quad N \to \infty, \quad (10.17)$$

and

$$S_n^{(l)} \sim \frac{R_0}{R_0 - 1}, \quad l = 0, 1, \quad M_1 \le n \le N, \quad R_0 > 1, \quad N \to \infty. \quad (10.18)$$

We note that there is a considerable overlap between the n-values for which the two approximations in (10.16) and (10.18) are valid. The results show that $S_n^{(l)}$ grows quickly with n from $S_1^{(l)} = 1$ toward the constant value $R_0/(R_0 - 1)$.

To derive the first of these approximations we use the approximation of $r_k^{(l)}$ given in (10.13). Inserting it into the definition of $S_n^{(l)}$ in (10.3) leads to the following approximation of $S_n^{(l)}$:

$$S_n^{(l)} = \sum_{k=1}^{n} r_k^{(l)} \sim \sum_{k=1}^{n} \frac{1}{\rho_k}, \quad l = 0, 1, \quad 1 \le n \le M_3, \quad R_0 > 1, \quad N \to \infty. \quad (10.19)$$

Here we apply the approximation of ρ_k given in (6.43). It is valid for n-values that are of smaller order than N. Thus we get

$$S_n^{(l)} \sim R_0\varphi(\beta_2) \sum_{k=1}^{n} \frac{1}{\varphi(y_2(k))}, \quad l = 0, 1, \quad 1 \le n, \quad n = \mathrm{o}(N), \quad R_0 > 1, \quad N \to \infty. \quad (10.20)$$

At this point we make use of the approximation of a sum of reciprocals of normal densities derived in Chap. 5 and given by (5.96), with F^* determined from (5.100). The reason for choosing this particular expression for F^* is that both 0 and n lie in the left far-end tail of the distribution with mean $\bar{\mu}_2$ and standard deviation $\bar{\sigma}_2$. To confirm this, we evaluate $y_2(n)$ and $z_2(n)$:

$$y_2(n) = \frac{n - \bar{\mu}_2}{\bar{\sigma}_2} = \frac{n - N \log R_0}{\sqrt{N}} = \frac{n}{\sqrt{N}} - \sqrt{N} \log R_0, \quad (10.21)$$

and

$$z_2(n) = \frac{y_2(n)}{\bar{\sigma}_2} = \frac{n}{N} - \log R_0. \quad (10.22)$$

Any nonnegative n that is of smaller order then N will clearly give $n/\sqrt{N} = o(\sqrt{N})$. Therefore, the first term in the last expression for $y_2(n)$ is of smaller order than the second one. This implies that $y_2(n) \to -\infty$ as $N \to \infty$ if $n = o(N)$. Furthermore, we get $z_2(n)$ of the order of one under the same conditions. This means that $n = o(N)$ lies in the left far-end tail of the distribution with mean $\bar{\mu}_2$ and standard deviation $\bar{\sigma}_2$. Thus, applying (5.96) and (5.100) gives the following approximation of $S_n^{(1)}$:

$$S_n^{(l)} \sim R_0 \varphi(\beta_2) \left[\frac{\varphi^*(y_2(n))}{1 - \exp(-z_2(n))} - \frac{\varphi^*(y_2(0))}{1 - \exp(-z_2(0))} \right],$$

$$l = 0, 1, \quad 1 \leq n, \quad n = o(N), \quad R_0 > 1, \quad N \to \infty. \quad (10.23)$$

This expression is simplified by the observations that $\varphi^*(y) = 1/\varphi(y)$, $y_2(0) = -\beta_2$, $z_2(0) = -\log R_0$, and that $z_2(n) \sim -\log R_0$ for $n = o(N)$. Insertions of these expressions into (10.23) gives

$$S_n^{(l)} \sim R_0 \varphi(\beta_2) \left[\frac{1}{\varphi(y_2(n))} \frac{1}{1 - R_0} - \frac{1}{\varphi(\beta_2)} \frac{1}{1 - R_0} \right],$$

$$l = 0, 1, \quad 1 \leq n, \quad n = o(N), \quad R_0 > 1, \quad N \to \infty. \quad (10.24)$$

It is straightforward to show that this approximation is equal to the approximation for $S_n^{(l)}$ claimed in (10.16).

The approximation of $S_n^{(l)}$ in (10.17) follows readily from the approximation in (10.16) by application of the relation (6.45), which shows that $\varphi(y_2(n))/\varphi(\beta_2) \sim R_0^n$ when n is of smaller order than $\sqrt{N}$, as it is when $1 \leq n \leq M_1 = \left[(N/R_0)^{1/4} \right]$.

To derive the approximation of $S_n^{(l)}$ in (10.18), we note first that

$$S_n^{(l)} = S_{M_1}^{(l)} + \sum_{k=M_1+1}^{n} r_k^{(l)}, \quad l = 0, 1, \quad M_1 \leq n \leq N. \quad (10.25)$$

Here, (10.17) gives an asymptotic approximation of the first term. We get

$$S_{M_1}^{(l)} \sim \frac{R_0}{R_0 - 1} - \frac{1}{R_0 - 1} \frac{1}{R_0^{M_1 - 1}}, \quad l = 0, 1, \quad N \to \infty. \quad (10.26)$$

The second term in the right-hand side of this relation is exponentially small, since M_1 is of the order of $N^{1/4}$. Furthermore, we apply the bound $r_k^{(l)} \leq r_{M_1}^{(l)}$ in (10.14), which is valid for $M_1 \leq k \leq N$, and use the asymptotic approximation of $r_{M_1}^{(l)}$ in (10.15) to get

$$\sum_{k=M_1+1}^{n} r_k^{(l)} \leq \sum_{k=M_1+1}^{n} r_{M_1}^{(l)} < N r_{M_1}^{(l)} \sim \frac{N}{R_0^{M_1-1}}, \quad l=0,1, \quad M_1 \leq n \leq N, \quad N \to \infty.$$
(10.27)

This asymptotic approximation is again seen to be exponentially small. Thus, the approximation of $S_n^{(l)}$ given in (10.18) has been established.

The third step is to derive the approximation of $\tilde{p}_1^{(l)}$ given in (10.10). The derivation is based on the expression (10.5) for $p_1^{(l)}$ for the two l-values. We show below that the denominators in these expressions for $\tilde{p}_1^{(l)}$ have the following asymptotic approximations:

$$\sum_{n=1}^{N} \pi_n S_n^{(l)} \sim \frac{R_0}{R_0-1} \sum_{n=1}^{N} \pi_n, \quad l=0,1, \quad R_0 > 1, \quad N \to \infty.$$
(10.28)

Inserting these approximations into (10.5), and using the relation (3.7) that expresses the sum $\sum_{n=1}^{N} \pi_n$ as $1/p_1^{(0)}$, we get

$$\tilde{p}_1^{(l)} \sim \frac{R_0-1}{R_0} p_1^{(0)}, \quad l=0,1, \quad R_0 > 1, \quad N \to \infty.$$
(10.29)

Insertion of the asymptotic approximation (9.5) of $p_1^{(0)}$ establishes the approximations (10.10) of $\tilde{p}_1^{(l)}$ for both values of l.

To prove (10.28), we write

$$A = \sum_{n=1}^{N} \pi_n, \quad A_1 = \sum_{n=1}^{M_1} \pi_n, \quad A_2 = \sum_{n=M_1+1}^{N} \pi_n,$$
(10.30)

and note that
$$A = A_1 + A_2.$$
(10.31)

We show that A_1/A is exponentially small. We use the result that π_n is smaller than $\max(1, \pi_{M_1}) = \pi_{M_1}$ for $1 \leq n \leq M_1$. We get therefore

$$A_1 = \sum_{n=1}^{M_1} \pi_n < M_1 \pi_{M_1} = \rho_{M_1}.$$
(10.32)

Moreover, we note that $A = 1/p_1^{(0)}$, and use the approximation $p_1^{(0)} \sim (R_0 - 1)$ $\varphi(\beta_1)\sqrt{N}$ given in (9.5), and the approximation $\rho_n \sim R_0^{n-1}$ given in (6.44) for $n = o(\sqrt{N})$. We get

$$\frac{A_1}{A} < p_1^{(0)} \rho_{M_1} \sim \frac{R_0 - 1}{R_0} \varphi(\beta_1)\sqrt{N} R_0^{M_1}, \quad N \to \infty. \tag{10.33}$$

Among the factors in the asympotic approximation of the upper bound of A_1/A, we find that $\varphi(\beta_1)$ is exponentially small, and that $R_0^{M_1}$ is exponentially large. To determine the behavior of the product, we insert the expression (6.35) for β_1 to find that

$$\varphi(\beta_1) R_0^{M_1} = \frac{1}{\sqrt{2\pi}} \exp\left(-N\left(\log R_0 - \frac{R_0 - 1}{R_0}\right) + M_1 \log R_0\right). \tag{10.34}$$

The exponent approaches $-\infty$ as $N \to \infty$, since M_1 is of the order of $N^{1/4}$. The product of the two important functions is therefore asymptotically small. Thus, we conclude that A_1/A is asymptotically small, and that $A \sim A_2$ as $N \to \infty$.

We introduce also the sums

$$B^{(l)} = \sum_{n=1}^{N} \pi_n S_n^{(l)}, \quad B_1^{(l)} = \sum_{n=1}^{M_1} \pi_n S_n^{(l)}, \quad B_2^{(l)} = \sum_{n=M_1+1}^{N} \pi_n S_n^{(l)}, \quad l = 0, 1, \tag{10.35}$$

where the first one is the sum of the other two:

$$B^{(l)} = B_1^{(l)} + B_2^{(l)}, \quad l = 0, 1. \tag{10.36}$$

By using the approximations (10.17) and (10.18) of $S_n^{(l)}$ we get

$$B_1^{(l)} < \frac{R_0}{R_0 - 1} A_1, \quad l = 0, 1, \quad R_0 > 1, \quad N \to \infty, \tag{10.37}$$

$$B_2^{(l)} \sim \frac{R_0}{R_0 - 1} A_2, \quad l = 0, 1, \quad R_0 > 1, \quad N \to \infty. \tag{10.38}$$

It follows that $B_1^{(l)}/B_2^{(l)}$ is exponentially small since A_1/A is, and that therefore $B^{(l)} \sim B_2^{(l)}$:

$$\frac{B_1^{(l)}}{B_2^{(l)}} < \frac{A_1}{A_2} \sim \frac{A_1}{A}, \quad l = 0, 1, \quad R_0 > 1, \quad N \to \infty. \tag{10.39}$$

We conclude that

$$B^{(l)} \sim B_2^{(l)} \sim \frac{R_0}{R_0 - 1} A_2 \sim \frac{R_0}{R_0 - 1} A, \quad l = 0, 1, \quad R_0 > 1, \quad N \to \infty. \tag{10.40}$$

Thus, (10.28) has been established. This proves that the approximations of $\tilde{p}_1^{(l)}$ given in (10.10) hold.

We have now acquired approximations of all three factors that determine $\tilde{p}_n^{(l)}$ in the right-hand sides of relations (10.2), namely π_n, $S_n^{(l)}$, and $\tilde{p}_1^{(l)}$ for $l = 0$ and $l = 1$. It is a pleasant exercise to insert the relevant approximations of these factors into (10.2), and thus verify that the approximations of the left tails of the distributions $\tilde{p}^{(l)}$ given in (10.7) for the two l-values hold. We use the approximation (8.4) of π_n, the approximations (10.16) of $S_n^{(l)}$, and the approximations (10.10) of $\tilde{p}_1^{(l)}$.

We remark that the same approach can be used to verify the approximation of the body and the near-end tails of the distributions $\tilde{p}^{(l)}$ given in (10.6) for the two l-values. In this case, we use the approximation (8.3) of π_n, the approximation (10.18) of $S_n^{(l)}$, and, as in the previous case, the approximation (10.10) of $\tilde{p}_1^{(l)}$.

It is illuminating to find that asymptotic approximations of the bodies and near-end tails of the two distributions $\tilde{\mathbf{p}}^{(0)}$ and $\tilde{\mathbf{p}}^{(1)}$ in the parameter region $R_0 > 1$ coincide with the asymptotic approximations of bodies and near-end tails of the original distributions $\mathbf{p}^{(0)}$ and $\mathbf{p}^{(1)}$, but that clear differences are found when we investigate the important left tails.

10.2   Approximation of $\Psi\left(\mathbf{p}^{(1)}\right)$ when R_0 is Distinctly Below One

We give here one-term asymptotic approximations of the components of the probability vector $\Psi\left(\mathbf{p}^{(1)}\right)$ in the parameter region where R_0 is distinctly below one. The surprising result is that these approximations agree with the one-term asymptotic approximations of the stationary distribution $\mathbf{p}^{(1)}$. Thus, whatever changes that the map Ψ causes when it is applied to the vector $\mathbf{p}^{(1)}$, they are not discernible when we compare the one-term asymptotic approximation of the vector $\mathbf{p}^{(1)}$ with the one-term asymptotic approximation of its image $\tilde{\mathbf{p}}^{(1)}$ under Ψ. We give also asymptotic approximations of the distribution function, defined by

$$\tilde{F}^{(1)}(n) = \sum_{k=1}^{n} \tilde{p}_k^{(1)}, \quad n = 1, 2, \ldots, N. \tag{10.41}$$

The results are summarized in the following theorem:

Theorem 10.2 *The distribution* $\tilde{\mathbf{p}}^{(1)} = \Psi\left(\mathbf{p}^{(1)}\right)$ *for the SIS model is approximated as follows when $R_0 < 1$ as $N \to \infty$:*

$$\tilde{p}_n^{(1)} \sim \frac{1 - R_0}{R_0} \frac{\varphi(y_2(n))}{\varphi(\beta_2)}, \quad R_0 < 1, \quad n = o(N), \quad 1 \le n, \quad N \to \infty, \tag{10.42}$$

$$\tilde{p}_n^{(1)} \sim (1 - R_0)R_0^{n-1}, \qquad R_0 < 1, \quad n = o(\sqrt{N}), \quad 1 \le n, \quad N \to \infty, \qquad (10.43)$$

$$\tilde{p}_1^{(1)} \sim 1 - R_0, \qquad R_0 < 1, \quad N \to \infty, \qquad (10.44)$$

$$\tilde{F}^{(1)}(n) \sim 1 - \frac{\varphi(y_2(n))}{\varphi(\beta_2)}, \qquad R_0 < 1, \quad n = o(N), \qquad 1 \le n, \quad N \to \infty, \qquad (10.45)$$

$$\tilde{F}^{(1)}(n) \sim 1 - R_0^n, \qquad R_0 < 1, \quad n = o(\sqrt{N}), \quad 1 \le n, \quad N \to \infty. \qquad (10.46)$$

The derivation of the basic result in (10.42) follows the same four steps as in the preceding section. In the first step we determine an approximation of the ratio $r_k^{(1)}$ defined in (10.4), while the second step leads us to an approximation of the sum $S_n^{(1)}$ in (10.3), and the third step gives the approximation of $\tilde{p}_1^{(1)}$ given in (10.44), using (10.5). The fourth step leads to the approximation of $\tilde{p}_n^{(1)}$ given in (10.42), using (10.2).

The approximation in the left-most tail given in (10.43) follows from the approximation in (10.42) by application of the asymptotic approximation $\varphi(y_2(n))/\varphi(\beta_2) \sim R_0^n$ for $n = o(\sqrt{N})$ in (6.45). The approximation of the distribution function $\tilde{F}^{(1)}(n)$ in (10.45) follows from (10.42) in exactly the same way as (7.10) follows from (7.7). The approximation of the distribution function in (10.46) follows from the approximation in (10.45) after application of the asymptotic approximation for $\varphi(y_2(n))/\varphi(\beta_2)$ in (6.45).

We proceed to derive the result in (10.42). We put $M_2 = [(N/R_0)^{3/4}]$ as in (6.61). The first step in the derivation will lead us to the following approximation and bound for $r_k^{(1)}$:

$$r_k^{(1)} \sim 1, \quad R_0 < 1, \quad k = 1,2,\ldots,M_2, \quad N \to \infty, \qquad (10.47)$$

$$r_k^{(1)} < 1, \quad R_0 < 1, \quad k = M_2, M_2 + 1, \ldots, N, \quad N \to \infty. \qquad (10.48)$$

The ratio $r_k^{(1)}$ associated with the distribution $\mathbf{p}^{(1)}$ is found to be as follows:

$$r_k^{(1)} = \frac{1 - \sum_{j=1}^{k-1} p_j^{(1)}}{\rho_k} = \frac{1 - F^{(1)}(k-1)}{\rho_k}, \quad k = 1,2,\ldots,N, \qquad (10.49)$$

where $F^{(1)}$ defined in (7.6) denotes the distribution function for the stationary distribution $\mathbf{p}^{(1)}$. By using the approximation of $F^{(1)}(k)$ given in (7.10) and the approximation of ρ_k in (6.43) we get

$$r_k^{(1)} \sim \frac{R_0\varphi(y_2(k-1))}{\varphi(y_2(k))} \sim 1, \quad R_0 < 1, \quad k = o(N), \quad N \to \infty. \qquad (10.50)$$

The last approximation step follows since

$$\frac{\varphi(y_2(k-1))}{\varphi(y_2(k))} = \exp\left(-\log R_0 + \frac{2k-1}{2N}\right) \sim \frac{1}{R_0}, \quad k = o(N), \quad N \to \infty. \qquad (10.51)$$

Thus, (10.47) has been established. The bound given in (10.48) actually holds for all $k > 1$, since $r_1^{(1)} = 1$ and we have proved in Sect. 3.9 that $r_k^{(1)}$ decreases in k.

The results concerning the ratio $r_k^{(1)}$ in (10.47) and (10.48) lead immediately to the following approximation and bound of the sum $S_n^{(1)}$:

$$S_n^{(1)} \sim n, \quad R_0 < 1, \quad n = 1, 2, \ldots, M_2, \quad N \to \infty, \tag{10.52}$$

and

$$S_n^{(1)} < n, \quad R_0 < 1, \quad n = M_2, M_2 + 1, \ldots, N. \tag{10.53}$$

Our next step is to derive the approximation of the probability $\tilde{p}_1^{(1)}$ given in (10.44). We use the relation (10.5). We introduce the quantities A, A_1, A_2, B_2 as follows:

$$A = \sum_{n=1}^{N} \pi_n S_n^{(1)}, \quad A_1 = \sum_{n=1}^{M_2} \pi_n S_n^{(1)}, \quad A_2 = \sum_{n=M_2+1}^{N} \pi_n S_n^{(1)}, \quad B_2 = \sum_{n=M_2+1}^{N} n\pi_n. \tag{10.54}$$

These quantities are related as follows:

$$A = A_1 + A_2, \quad A_2 < B_2. \tag{10.55}$$

We show that

$$A_1 \sim \frac{1}{1 - R_0}, \quad R_0 < 1, \quad N \to \infty, \tag{10.56}$$

and that

$$\frac{B_2}{A_1} = o(1), \quad N \to \infty. \tag{10.57}$$

Insertion into (10.5) shows that the approximation (10.44) of the probability $\tilde{p}_1^{(1)}$ holds.

In order to derive the approximation (10.56) of A_1, we note first that the approximation (10.52) of $S_n^{(1)}$, and the relation $\rho_n = n\pi_n$ imply that

$$A_1 \sim \sum_{n=1}^{M_2} n\pi_n = \sum_{n=1}^{M_2} \rho_n = S_1, \quad N \to \infty, \tag{10.58}$$

and

$$B_2 = \sum_{n=M_2+1}^{N} \rho_n = S_2, \tag{10.59}$$

where S_1 and S_2 are defined by (6.80) in Sect. 6.3.2.

The results in (10.56) and (10.57) follow from the results established in Sect. 6.3.2 that $S_1 \sim 1/(1 - R_0)$ and that S_2 is exponentially small. We conclude, using (10.5), that the approximation of $\tilde{p}_1^{(1)}$ in (10.44) holds.

The final step can now be taken. We insert the approximation of π_n from (8.4), the approximation of $S_n^{(1)}$ from (10.52), and the approximation of $\tilde{p}_1^{(1)}$ from (10.44) into (10.2). This completes the derivation of the approximation of $\tilde{p}_n^{(1)}$ in (10.42).

We conclude with an important observation. Let us determine the one-term asymptotic approximation of $\Psi^n\left(\mathbf{p}^{(1)}\right)$ for any n larger than or equal to one. It is clear from the above result that this approximation is independent of n and equal to the asymptotic approximation of $\mathbf{p}^{(1)}$. This finding has important implications in the next chapter, where we study the quasi-stationary distribution.

We remark that a similar study of the probability vector $\Psi\left(\mathbf{p}^{(0)}\right)$ for $R_0 < 1$ has not met with success. We leave it as an open problem to investigate this distribution. However, it will probably not give any additional useful information about the most important descriptor of the SIS model, namely the quasi-stationary distribution.

10.3 Approximation of $\Psi\left(\mathbf{p}^{(1)}\right)$ in the Transition Region

This section is devoted to a study of the probability vector $\tilde{\mathbf{p}}^{(1)} = \Psi\left(\mathbf{p}^{(1)}\right)$ in the transition region. Approximation of this distribution is more of a challenge in the transition region than in the parameter regions where R_0 is distinctly different from one. The approximation that we derive in this section is not claimed to be asymptotic. It is left as an open problem to determine an asymptotic approximation of this distribution.

To describe the approximation of this section, we define three functions $\tilde{\rho}_1$, H_2, and R_1 as follows:

$$\tilde{\rho}_1(y) = y + \frac{1}{H_1(y)}, \tag{10.60}$$

$$H_2(y) = \frac{1}{y + 1/H_1(y)} \int_{-1/H_1(y)}^{y} H_1(t)\,dt, \tag{10.61}$$

$$R_1(N,y) = 1 + \tilde{\rho}_1(y)\frac{1}{\sqrt{N}}. \tag{10.62}$$

We prove the following properties of $\tilde{\rho}_1$ and H_2:

$$0 < \tilde{\rho}_1(y), \quad y \in \mathfrak{R}, \tag{10.63}$$

$$0 < H_2(y) < H_1(y), \quad y \in \mathfrak{R}. \tag{10.64}$$

The inequality $0 < \tilde{\rho}_1(y)$ in (10.63) follows from the definition of $\tilde{\rho}_1(y)$ for $y \geq 0$ since $H_1(y)$ is positive for all y. For $y < 0$, the same inequality follows since then $H_1(y) < -1/y$, by (5.45).

We note that $R_1(N,\rho)$ resembles R_0, since it is equal to one plus a function of ρ divided by $\sqrt{N}$. However, it differs from R_0 by being larger than one for both positive and negative values of ρ, while we recall that $R_0 - 1 = \rho/\sqrt{N}$ has the same sign as ρ. The function value $H_2(y)$ is seen from the definition of H_2 in (10.61) to be equal to the average of the function $H_1(t)$ over the interval from $-1/H_1(y)$ to y. The length of this interval is $\tilde{\rho}_1(y)$. The lower limit of integration is smaller than the upper limit, since $\tilde{\rho}_1(x) > 0$ for all x. The function value $H_2(y)$ is therefore always positive. We note also that $H_2(y) < H_1(y)$. The reason for this is that H_1 is a strictly increasing function. Therefore, the integrand $H_1(t)$ in (10.61) is smaller than $H_1(y)$. Replacing it by this upper bound, we find that $H_2(y) < H_1(y)$.

The main results of this section are summarized in the following theorem:

Theorem 10.3 *The distribution $\tilde{\mathbf{p}}^{(1)} = \Psi\left(\mathbf{p}^{(1)}\right)$ for the SIS model is approximated as follows in the transition region:*

$$\tilde{p}_n^{(1)} \approx \frac{1}{\varphi(\rho)\tilde{\rho}_1(\rho)H_2(\rho)}\frac{1}{n}\left(1-\frac{1}{[R_1(N,\rho)]^n}\right)\varphi(y_3(n)),$$

$$\rho = O(1), \quad 1 \leq n, \quad n = o(N), \quad N \to \infty, \quad (10.65)$$

$$\tilde{p}_n^{(1)} \approx \frac{1}{H_2(\rho)}\frac{1}{\sqrt{N}}, \quad \rho = O(1), \quad 1 \leq n, \quad n = o(\sqrt{N}), \quad N \to \infty. \quad (10.66)$$

$$\tilde{p}_1^{(1)} \approx \frac{1}{H_2(\rho)}\frac{1}{\sqrt{N}}, \quad \rho = O(1), \quad N \to \infty. \quad (10.67)$$

We note the similarity between the approximation of the left-most tail of the distribution $\tilde{p}_n^{(1)}$ in (10.66) and the approximation $1/(H_1(\rho)\sqrt{N})$ of the left-most tail of of $p_n^{(1)}$ in (7.27).

The second approximation in the above theorem is a special case of the first one, valid for a shorter range of n-values. The third approximation is really contained in the second one. We exhibit it for emphasis, since it will be derived before (10.65) is.

We show first that the second approximation can be derived from the first one, and leave the derivation of (10.65) to the end. By using the definition $R_1(N,\rho) = 1 + \tilde{\rho}_1(\rho)/\sqrt{N}$, we find that

$$\log[R_1(N,\rho)]^n = n\log\left(1+\frac{\tilde{\rho}_1(\rho)}{\sqrt{N}}\right) \sim \frac{n\tilde{\rho}_1(\rho)}{\sqrt{N}}, \quad \rho = O(1), \quad N \to \infty. \quad (10.68)$$

It follows that

$$[R_1(N,\rho)]^n \sim \exp\left(\frac{n\tilde{\rho}_1(\rho)}{\sqrt{N}}\right) \sim 1+\frac{n\tilde{\rho}_1(\rho)}{\sqrt{N}}, \quad \rho = O(1), \quad n = o(\sqrt{N}), \quad N \to \infty.$$

$$(10.69)$$

We get therefore

$$\frac{1}{n}\left(1 - \frac{1}{[R_1(N,\rho)]^n}\right) \sim \frac{\tilde{p}_1(\rho)}{\sqrt{N}}, \quad \rho = O(1), \quad n = o(\sqrt{N}), \quad N \to \infty. \quad (10.70)$$

Furthermore,

$$\frac{\varphi(y_3(n))}{\varphi(\rho)} = \frac{\varphi(n/\sqrt{N} - \rho)}{\varphi(\rho)} = \exp\left(\rho\frac{n}{\sqrt{N}} - \frac{1}{2}\frac{n^2}{N}\right) \sim 1,$$

$$\rho = O(1), \quad n = o(\sqrt{N}), \quad N \to \infty. \quad (10.71)$$

By applying these approximations of $(1 - 1/[R_1(N,\rho)]^n)/n$ and of $\varphi(y_3(n))/\varphi(\rho)$ to (10.65), we find that (10.66) holds.

We turn now to the derivation of the approximation (10.65). The method to be used is the same as in the two preceding sections. Thus, in the first step we approximate $r_k^{(1)}$, using its definition in (10.4). After that, we establish an approximation of he sum $S_n^{(1)}$ defined in (10.3). The third step is to derive the approximation of $\tilde{p}_1^{(1)}$ given in (10.67), using the expression in (10.5). Finally, we insert the approximation of π_n given in (8.4), and the approximations just derived of $S_n^{(1)}$ and $\tilde{p}_1^{(1)}$ into the expression (10.2) for $\tilde{p}_n^{(1)}$ to confirm the result claimed in (10.65).

We next derive the following approximation of $r_k^{(1)}$:

$$r_k^{(1)} \sim \frac{1}{[R_1(N,\rho)]^{k-1}}, \quad \rho = O(1), \quad k = o(\sqrt{N}), \quad 1 \le k, \quad N \to \infty. \quad (10.72)$$

To this end, we consider the ratio $r_{k+1}^{(1)}/r_k^{(1)}$. We recall from Sect. 3.9 that $r_k^{(1)}$ is decreasing in k. It follows therefore that the ratio is smaller than one. We show that the ratio can be approximated as follows:

$$\frac{r_{k+1}^{(1)}}{r_k^{(1)}} \sim \frac{1}{R_1(N,\rho)}, \quad \rho = O(1), \quad k = o(\sqrt{N}), \quad 1 \le k, \quad N \to \infty, \quad (10.73)$$

independent of k. Clearly, this implies (10.72). By using the definition of $r_k^{(1)}$ in (10.4), we get

$$\frac{r_{k+1}^{(1)}}{r_k^{(1)}} = \frac{\rho_k}{\rho_{k+1}}\frac{1 - \sum_{j=1}^{k} p_j^{(1)}}{1 - \sum_{j=1}^{k-1} p_j^{(1)}} = \frac{1}{1 - k/N}\frac{1}{R_0}\left(1 - \frac{p_k^{(1)}}{1 - F^{(1)}(k-1)}\right),$$

$$\rho = O(1), \quad k = 1, 2, \ldots, N-1, \quad (10.74)$$

where we have used the expression (4.7) to write $\rho_{k+1}/\rho_k = (1 - k/N)R_0$.

In the last factor of this expression we use the approximations of $p_n^{(1)}$ and of $F^{(1)}(n)$ derived for the transition region in Chap. 7 and given in (7.27) and (7.29), respectively. They hold for $\rho = O(1)$ and $n = o(\sqrt{N})$. Thus we get

$$\frac{p_k^{(1)}}{1 - F^{(1)}(k-1)} \sim \frac{1}{H_1(\rho)} \frac{1}{\sqrt{N}},$$

$$\rho = O(1), \quad k = o(\sqrt{N}), \quad 1 \leq k, \quad N \to \infty. \quad (10.75)$$

Insertion into (10.74) gives

$$\frac{r_{k+1}^{(1)}}{r_k^{(1)}} \sim \frac{1}{1 + \rho/\sqrt{N}} \left(1 - \frac{1}{H_1(\rho)\sqrt{N}}\right) \sim \frac{1}{R_1(N,\rho)},$$

$$\rho = O(1), \quad k = o(\sqrt{N}), \quad 1 \leq k, \quad N \to \infty, \quad (10.76)$$

as claimed in (10.73). We conclude that the approximation of $r_k^{(1)}$ given in (10.72) holds.

The next step is simple. It follows that the sum $S_n^{(1)}$ can be approximated as follows:

$$S_n^{(1)} \sim \sum_{k=1}^{n} \frac{1}{[R_1(N,\rho)]^{k-1}} = \frac{R_1(N,\rho)}{R_1(N,\rho) - 1} \left(1 - \frac{1}{[R_1(N,\rho)]^n}\right),$$

$$\sim \frac{\sqrt{N}}{\tilde{\rho}_1(\rho)} \left(1 - \frac{1}{[R_1(N,\rho)]^n}\right), \quad \rho = O(1), \quad n = o(\sqrt{N}), \quad 1 \leq n, \quad N \to \infty.$$

$$(10.77)$$

Here we have used the fact that $R_1(N,\rho) \sim 1$ in the transition region.

Next we give a derivation of the approximation of the probability $\tilde{p}_1^{(1)}$ given in (10.67). We base the derivation on an approximation of $\sum_{n=1}^{N} \pi_n S_n^{(1)}$, which by (10.5) equals the inverse of $\tilde{p}_1^{(1)}$. In this step, we shall introduce two approximations that are known to be non-asymptotic. Both of them consist in extending known asymptotic approximations to larger n-values than those for which the approximations are known to be asymptotic. The first of these extensions concerns $S_n^{(1)}$. We shall use the approximation (10.77) for all n-values. This means that we use

$$S_n^{(1)} \approx \frac{\sqrt{N}}{\tilde{\rho}_1(\rho)} \left(1 - \frac{1}{[R_1(N,\rho)]^n}\right), \quad \rho = O(1), \quad n = 1,2,\ldots,N, \quad N \to \infty.$$

$$(10.78)$$

Using this approximation, we find that the important sum $\sum_{n=1}^{N} \pi_n S_n^{(1)}$ is approximated as follows:

$$\sum_{n=1}^{N} \pi_n S_n^{(1)} \approx \frac{\sqrt{N}}{\tilde{\rho}_1(\rho)} \left(\sum_{n=1}^{N} \pi_n(R_0) - \sum_{n=1}^{N} \frac{\pi_n(R_0)}{[R_1(N,\rho)]^n} \right), \quad \rho = O(1), \quad N \to \infty,$$
(10.79)

where we have written $\pi_n(R_0)$ instead of π_n to emphasize that π_n depends on R_0.

We recognize that the first of the two sums in the right-hand side of this expression is approximated by (8.14). Thus, we get

$$\sum_{n=1}^{N} \pi_n(R_0) \approx \frac{1}{2} \log N + H_0(\rho), \quad \rho = O(1), \quad N \to \infty. \tag{10.80}$$

We need also an approximation of the second sum. We are going to make use of the asymptotic approximation of π_n that is given by (8.5), namely $\pi_n \sim R_0^{n-1}/n$, which is valid for $n = o(\sqrt{N})$. By using this approximation, we get

$$\frac{\pi_n(R_0)}{[R_1(N,\rho)]^n} \sim \frac{1}{R_1(N,\rho)} \frac{1}{n} \left(\frac{R_0}{R_1(N,\rho)} \right)^{n-1} \sim \pi_n \left(\frac{R_0}{R_1(N,\rho)} \right),$$
$$\rho = O(1), \quad n = o(\sqrt{N}), \quad N \to \infty. \tag{10.81}$$

The second non-asymptotic approximation that we introduce is to use this approximation for all n-values. Proceeding, we get the following approximation of the second sum in (10.79):

$$\sum_{n=1}^{N} \frac{\pi_n(R_0)}{[R_1(N,\rho)]^n} \approx \frac{1}{R_1(N,\rho)} \sum_{n=1}^{N} \pi_n \left(\frac{R_0}{R_1(N,\rho)} \right)$$
$$\approx \frac{1}{2} \log N + H_0 \left(\rho \left(\frac{R_0}{R_1(N,\rho)} \right) \right),$$
$$\rho = O(1), \quad N \to \infty. \tag{10.82}$$

Here, we note that $\rho = \rho(R_0) = (R_0 - 1)\sqrt{N}$. In the above expression we need to evaluate $\rho(R_0/R_1(N,\rho))$. To prepare for this, we evaluate the ratio $R_0/R_1(N,\rho)$. We get

$$\frac{R_0}{R_1(N,\rho)} = \frac{1 + \rho/\sqrt{N}}{1 + (\rho + 1/H_1(\rho))/\sqrt{N}} \sim 1 - \frac{1}{H_1(\rho)} \frac{1}{\sqrt{N}}, \quad \rho = O(1), \quad N \to \infty. \tag{10.83}$$

This leads to the approximation

$$\rho \left(\frac{R_0}{R_1(N,\rho)} \right) = \left(\frac{R_0}{R_1(N,\rho)} - 1 \right) \sqrt{N} \sim -\frac{1}{H_1(\rho)}, \quad \rho = O(1), \quad N \to \infty. \tag{10.84}$$

The approximation of the sum in (10.82) can therefore be written

$$\sum_{n=1}^{N} \frac{\pi_n(R_0)}{[R_1(N,\rho)]^n} \approx \frac{1}{2}\log N + H_0\left(-\frac{1}{H_1(\rho)}\right), \quad \rho = O(1), \quad N \to \infty. \quad (10.85)$$

We insert the approximations in (10.80) and (10.85) into (10.79). This gives

$$\sum_{n=1}^{N} \pi_n S_n^{(1)} \approx \frac{\sqrt{N}}{\tilde{\rho}_1(\rho)}\left(\frac{1}{2}\log N + H_0(\rho) - \frac{1}{2}\log N - H_0\left(-\frac{1}{H_1(\rho)}\right)\right)$$

$$= \frac{\sqrt{N}}{\tilde{\rho}_1(\rho)}\int_{-1/H_1(\rho)}^{\rho} H_1(t)\,dt, \quad \rho = O(1), \quad N \to \infty. \quad (10.86)$$

By using the definitions of H_2 and $\tilde{\rho}_1$ we get

$$\sum_{n=1}^{N} \pi_n S_n^{(1)} \approx H_2(\rho)\sqrt{N}, \quad \rho = O(1), \quad N \to \infty. \quad (10.87)$$

The approximation $\tilde{p}_1^{(1)} \approx 1/(H_2(\rho)\sqrt{N})$ given in (10.67) follows from this result, since $\tilde{p}_1^{(1)} = 1/\sum_{n=1}^{N}\pi_n S_n^{(1)}$.

The final step in the derivation of the approximation of $\tilde{p}_n^{(1)}$ in (10.65) can now be taken. It consists of inserting the approximation of π_n from (8.4), the approximation of $S_n^{(1)}$ given in (10.77), and the approximation of $\tilde{p}_1^{(1)}$ in (10.67) into the expression $\tilde{p}_n^{(1)} = \pi_n S_n^{(1)}\tilde{p}_1^{(1)}$ given for $\tilde{p}_n^{(1)}$ in (10.2). In doing this, we recognize that π_n from (8.4) can be simplified as follows:

$$\pi_n \sim \frac{1}{nR_0}\frac{\varphi(y_2(n))}{\varphi(\beta_2)} \sim \frac{1}{n}\frac{\varphi(y_3(n))}{\varphi(\rho)}, \quad \rho = O(1), \quad n = o(N), \quad 1 \le n, \quad N \to \infty.$$
$$(10.88)$$

This completes the derivation of the approximation of $\tilde{p}_n^{(1)}$ in (10.65). Thus, all three approximations in Theorem 10.3 have been derived.

10.4 Approximation of $\Psi^i\left(\mathbf{p}^{(1)}\right)$ in the Transition Region

This section is used to derive approximations of the images of the probability vector $\mathbf{p}^{(1)}$ under the i-fold application of the map Ψ in the transition region. We use the notation $\tilde{\mathbf{p}}^{(i,1)} = \Psi^i\left(\mathbf{p}^{(1)}\right)$. Thus, the superscript used for the probability vector $\tilde{\mathbf{p}}^{(i,1)}$ contains two integers, of which the first one reminds us that this vector is the result of an i-fold application of Ψ, and the second one informs us that the original vector on which Ψ is applied is the vector $\mathbf{p}^{(1)}$. We note that the notation used in the previous section was $\tilde{\mathbf{p}}^{(1)} = \Psi\left(\mathbf{p}^{(1)}\right)$. Thus, the relation between old and new notation is shown by $\tilde{\mathbf{p}}^{(1,1)} = \tilde{\mathbf{p}}^{(1)}$.

We need some further notation to describe our results. We define recursively a sequence of functions H_i, starting from the function H_1:

$$H_i(y) = \frac{1}{y + 1/H_{i-1}(y)} \int_{-1/H_{i-1}(y)}^{y} H_1(t)\,dt, \quad i = 2,3,\ldots, \quad y \in \Re. \tag{10.89}$$

Thus, each $H_i(y)$ is the average of the function $H_1(t)$ over the interval from $-1/H_{i-1}(y)$ to y. We note that the function H_i is well defined, since the interval of integration is determined by the previously determined function H_{i-1}. We note also that the function H_2 defined by (10.61) belongs to this newly defined sequence of functions. Furthermore, it is useful to define a sequence of functions $\tilde{\rho}_i$ as follows:

$$\tilde{\rho}_i(y) = y + \frac{1}{H_i(y)}, \quad i = 1,2,\ldots,. \tag{10.90}$$

We note that this notation is consistent with the definition $\tilde{\rho}_1(y) = y + 1/H_1(y)$ in (10.60). Next we define a sequence of functions R_i by setting

$$R_i(N,y) = 1 + \frac{\tilde{\rho}_i(y)}{\sqrt{N}}, \quad i = 1,2,\ldots. \tag{10.91}$$

Clearly, the first element in this sequence coincides with the function $R_1(N,y)$ defined by (10.62). We define also

$$r_k^{(i,1)} = \frac{1 - \sum_{j=1}^{k-1} \tilde{p}_j^{(i-1,1)}}{\rho_k}, \quad k = 1,2,\ldots,N, \quad i = 1,2,\ldots, \tag{10.92}$$

and

$$S_n^{(i,1)} = \sum_{k=1}^{n} r_k^{(i,1)}, \quad n = 1,2,\ldots,N, \quad i = 1,2,\ldots. \tag{10.93}$$

Now we have $\tilde{\mathbf{p}}^{(i,1)} = \Psi\left(\tilde{\mathbf{p}}^{(i-1,1)}\right)$ for $i = 1,2,\ldots$, with the natural interpretation that $\tilde{\mathbf{p}}^{(0,1)} = \mathbf{p}^{(1)}$. The components of the probability vector $\tilde{\mathbf{p}}^{(i,1)}$ can therefore be written as follows:

$$\tilde{p}_n^{(i,1)} = \tilde{p}_1^{(i,1)} \pi_n S_n^{(i,1)}, \quad n = 1,2,\ldots,N, \quad i = 1,2,\ldots, \tag{10.94}$$

where the important probability $\tilde{p}_1^{(i,1)}$ is determined as follows:

$$\tilde{p}_1^{(i,1)} = \frac{1}{\sum_{n=1}^{N} \pi_n S_n^{(i,1)}}, \quad i = 1,2,\ldots. \tag{10.95}$$

The main result of this section is contained in the following theorem.

Theorem 10.4 *The distribution* $\tilde{\mathbf{p}}^{(i,1)} = \Psi^i\left(\mathbf{p}^{(1)}\right)$ *for the SIS model is approximated as follows in the transition region:*

$$\tilde{p}_n^{(i,1)} \approx \frac{1}{\varphi(\rho)\tilde{p}_i(\rho)H_{i+1}(\rho)}\frac{1}{n}\left(1 - \frac{1}{[R_i(N,\rho)]^n}\right)\varphi(y_3(n)),$$

$$\rho = O(1), \quad n = o(N), \quad 1 \le n, \quad i = 1,2,3,\ldots, \quad N \to \infty, \quad (10.96)$$

$$\tilde{p}_n^{(i,1)} \approx \frac{1}{H_{i+1}(\rho)\sqrt{N}},$$

$$\rho = O(1), \quad n = o(\sqrt{N}), \quad 1 \le n, \quad i = 1,2,3,\ldots, \quad N \to \infty, \quad (10.97)$$

$$\tilde{p}_1^{(i,1)} \approx \frac{1}{H_{i+1}(\rho)\sqrt{N}}, \quad \rho = O(1), \quad i = 1,2,3,\ldots, \quad N \to \infty. \quad (10.98)$$

We do not claim that these approximations are asymptotic. We derive the results in this theorem by using induction on i. Clearly, the first induction step has already been carried out, since the above results were shown to hold for $i = 1$ in Theorem 10.3.

As in the previous section, we note that the second approximation in the theorem is a special case of the first one, valid for a shorter range of n-values. Also, the third approximation is contained in the second one. However, we show it separately, since indeed it will be derived before the more generally valid approximation in (10.96).

We show first that the second approximation can be derived from the first one. The arguments are similar to those given in the previous section. By using the definition $R_i(N,\rho) = 1 + \tilde{p}_i(\rho)/\sqrt{N}$ we get

$$\log[R_i(N,\rho)]^n = n\log\left(1 + \frac{\tilde{p}_i(\rho)}{\sqrt{N}}\right) \sim \frac{n\tilde{p}_i(\rho)}{\sqrt{N}}, \quad N \to \infty. \quad (10.99)$$

It follows that

$$[R_i(N,\rho)]^n \sim \exp\left(\frac{n\tilde{p}_i(\rho)}{\sqrt{N}}\right) \sim 1 + \frac{n\tilde{p}_i(\rho)}{\sqrt{N}}, \quad n = o(\sqrt{N}), \quad N \to \infty. \quad (10.100)$$

Hence, we get

$$\frac{1}{n}\left(1 - \frac{1}{[R_i(N,\rho)]^n}\right) \sim \frac{\tilde{p}_i(\rho)}{\sqrt{N}}, \quad n = o(\sqrt{N}), \quad N \to \infty. \quad (10.101)$$

By applying this approximation and the approximation $\varphi(y_3(n))/\varphi(\rho) \sim 1$ in (10.71) to (10.96), we find that (10.97) holds.

We proceed to derive (10.96). We assume that (10.96) holds for $i = j$, with $j \geq 1$, and show that it then holds also for $i = j + 1$. The arguments are similar to those given in the previous section.

The ratio of adjacent values of $r_k^{(j+1,1)}$ equals

$$
\frac{r_{k+1}^{(j+1,1)}}{r_k^{(j+1,1)}} = \frac{\rho_k}{\rho_{k+1}} \frac{1 - \sum_{l=1}^{k} \tilde{p}_l^{(j,1)}}{1 - \sum_{l=1}^{k-1} \tilde{p}_l^{(j,1)}} = \frac{1}{(1 - k/N)R_0} \left(1 - \frac{\tilde{p}_k^{(j,1)}}{1 - \sum_{l=1}^{k-1} \tilde{p}_l^{(j,1)}} \right),
$$

$$
k = 1, 2, \ldots, N - 1, \quad j = 1, 2, \ldots. \quad (10.102)
$$

The approximation $\tilde{p}_k^{(j,1)} \approx 1/(H_{j+1}(\rho)\sqrt{N})$ for $k = o(\sqrt{N})$ is valid according to the induction hypothesis. By applying it, we get

$$
\frac{r_{k+1}^{(j+1,1)}}{r_k^{(j+1,1)}} \approx \frac{1}{R_0} \left(1 - \frac{1}{H_{j+1}(\rho)\sqrt{N}} \right)
$$

$$
\sim \frac{1}{(1 + \rho/\sqrt{N})(1 + 1/(H_{j+1}(\rho)\sqrt{N})} \sim \frac{1}{R_{j+1}(N,\rho)},
$$

$$
\rho = O(1), \ k = o(\sqrt{N}), \ 1 \leq k, \ j = 1, 2, \ldots, \ N \to \infty. \quad (10.103)
$$

Recalling that $r_1^{(j+1,1)} = 1$, we get

$$
r_k^{(j+1,1)} \approx \frac{1}{[R_{j+1}(N,\rho)]^{k-1}},
$$

$$
\rho = O(1), \quad k = o(\sqrt{N}), \quad 1 \leq k, \quad j = 1, 2, \ldots, \quad N \to \infty. \quad (10.104)
$$

From this approximation we get

$$
S_n^{(j+1,1)} = \sum_{k=1}^{n} r_k^{(j+1,1)} \approx \sum_{k=1}^{n} \frac{1}{[R_{j+1}(N,\rho)]^{k-1}}
$$

$$
= \frac{R_{j+1}(N,\rho)}{R_{j+1}(N,\rho) - 1} \left(1 - \frac{1}{[R_{j+1}(N,\rho)]^n} \right) \sim \frac{\sqrt{N}}{\tilde{\rho}_{j+1}(\rho)} \left(1 - \frac{1}{[R_{j+1}(N,\rho)]^n} \right),
$$

$$
\rho = O(1), \quad n = o(\sqrt{N}), \quad 1 \leq n, \quad j = 1, 2, \ldots, \quad N \to \infty. \quad (10.105)
$$

At this point we adopt the non-asymptotic approximation that this expression for $S_n^{(j+1,1)}$ holds for all values of n from 1 to N. This is a close parallel to the way in which the approximation of $S_n^{(1)}$ in the previous section is extended from being valid only for n-values that satisfy $n = o(\sqrt{N})$ to all n-values. Thus, we shall use the following approximation of $S_n^{(j+1,1)}$:

$$S_n^{(j+1,1)} \approx \frac{\sqrt{N}}{\tilde{p}_{j+1}(\rho)}\left(1 - \frac{1}{[R_{j+1}(N,\rho)]^n}\right),$$

$$\rho = O(1), \quad n = 1,2,\ldots,N, \quad j = 1,2,\ldots, \quad N \to \infty. \quad (10.106)$$

We proceed to consider the sum $\sum_{n=1}^N \pi_n S_n^{(j+1,1)}$. By using (10.95) we find that it equals $1/p_1^{(j+1,1)}$. We use the adopted approximation of $S_n^{(j+1,1)}$ in (10.106) to get

$$\sum_{n=1}^N \pi_n S_n^{(j+1,1)} \approx \frac{\sqrt{N}}{\tilde{p}_{j+1}(\rho)}\left(\sum_{n=1}^N \pi_n(R_0) - \sum_{n=1}^N \frac{\pi_n(R_0)}{[R_{j+1}(N,\rho)]^n}\right). \quad (10.107)$$

Proceeding as in the previous section, we find that the first sum in the right-hand side is approximated by $(1/2)\log N + H_0(\rho)$, and the second sum by $(1/2)\log N + H_0(-1/H_{j+1}(\rho))$. Thus,

$$\sum_{n=1}^N \pi_n S_n^{(j+1,1)} \approx \frac{\sqrt{N}}{\tilde{p}_{j+1}(\rho)}\int_{-1/H_{j+1}(\rho)}^{\rho} H_1(t)\,dt \sim \sqrt{N}H_{j+2}(\rho), \quad \rho = O(1), \quad N \to \infty.$$

$$(10.108)$$

This shows that the probability $\tilde{p}_1^{(j+1,1)}$ satisfies (10.98). Finally, insertion of this approximation of $\tilde{p}_1^{(j+1,1)}$, the approximation (10.88) of π_n, and the approximation (10.106) of $S_n^{(j+1,1)}$ into the expression $\tilde{p}_n^{(j+1,1)} = \tilde{p}_1^{(j+1,1)}\pi_n S_n^{(j+1,1)}$ given in (10.94) shows that $\tilde{p}_n^{(j+1,1)}$ satisfies (10.96). This concludes the inductive proof of (10.96).

Chapter 11
Approximation of the Quasi-stationary Distribution q of the SIS Model

The quasi-stationary distribution $\mathbf{q}$ has a central position in our study of the stochastic SIS model. We give approximations of this distribution in each of the three parameter regions in this chapter. As has been mentioned before, we actually work with two kinds of approximation of quite different type. The first type of approximation is a stationary distribution of some related process. Here we can identify the stationary distributions $\mathbf{p}^{(1)}$ and $\mathbf{p}^{(0)}$ of the two auxiliary processes discussed in Sect. 3.2 and Chap. 4, and for which approximations (of the second type) are derived in Chaps. 7 and 9. These auxiliary processes are both birth–death processes. In addition, we note that the image of arbitrary discrete distributions on $\{1, 2, \ldots, N\}$, under the map Ψ, as defined by Ferrari et al. [27], are actually stationary distributions of associated Markov Chains. We have found that the images under Ψ of the two stationary distributions $\mathbf{p}^{(1)}$ and $\mathbf{p}^{(0)}$ are particularly useful for our purposes. We shall also use the images $\Psi^i\left(\mathbf{p}^{(1)}\right)$ of the stationary distribution $\mathbf{p}^{(1)}$ after applying the map Ψ i times. The second type of approximation that we make use of is an analytical approximation of one of the approximations of the first type. It is desirable that this second type of approximation is asymptotic as $N \to \infty$, but we have not been able to find approximations that have this property in all cases.

11.1 Approximation of the Quasi-stationary Distribution q when R_0 is Distinctly Above One

We give four approximations of the quasi-stationary distribution $\mathbf{q}$ when R_0 is distinctly above the value one, one in the body and near-end tails, and three in the left tail. In addition, we give an approximation of the important probability q_1. The results are summarized in the following theorem:

Theorem 11.1 *The quasi-stationary distribution* $\mathbf{q}$ *of the SIS model is approximated as follows when* $R_0 > 1$ *as* $N \to \infty$:

I. Nåsell, *Extinction and Quasi-stationarity in the Stochastic Logistic SIS Model*,
Lecture Notes in Mathematics 2022, DOI 10.1007/978-3-642-20530-9_11,
© Springer-Verlag Berlin Heidelberg 2011

$$q_n \sim \frac{1}{\bar{\sigma}_1} \varphi(y_1(n)), \quad R_0 > 1, \quad y_1(n) = o(\sqrt{N}), \quad N \to \infty, \qquad (11.1)$$

$$q_n \sim \frac{R_0 - 1}{R_0} \varphi(\beta_1) \sqrt{N} \frac{1}{n} \left(\frac{\varphi(y_2(n))}{\varphi(\beta_2)} - 1 \right),$$
$$R_0 > 1, \quad n = o(N), \quad 1 \leq n, \quad N \to \infty, \quad (11.2)$$

$$q_n \sim \frac{R_0 - 1}{R_0} \varphi(\beta_1) \sqrt{N} \frac{1}{n} \frac{\varphi(y_2(n))}{\varphi(\beta_2)}, \quad R_0 > 1, \quad n = o(N), \quad n > M_1, \quad N \to \infty,$$
$$(11.3)$$

$$q_n \sim \frac{R_0 - 1}{R_0} \varphi(\beta_1) \sqrt{N} \frac{R_0^n - 1}{n}, \quad R_0 > 1, \quad n = o(\sqrt{N}), \quad 1 \leq n, \quad N \to \infty, \quad (11.4)$$

$$q_1 \sim \frac{(R_0 - 1)^2}{R_0} \varphi(\beta_1) \sqrt{N}, \quad R_0 > 1, \quad N \to \infty. \qquad (11.5)$$

Here, $M_1 = \left[\bar{\sigma}_1^{1/2} \right]$ is defined in (6.61).

Proof. The results in (11.1)-(11.5) follow directly from the corresponding results for the probability vectors $\tilde{\mathbf{p}}^{(1)}$ and $\tilde{\mathbf{p}}^{(0)}$ given in (10.6)-(10.10). The reason is that the quasi-stationary distribution **q** is stochastically bounded between $\tilde{\mathbf{p}}^{(0)}$ and $\tilde{\mathbf{p}}^{(1)}$. Indeed, it follows from (3.106) and the fact that likelihood ratio ordering implies stochastic ordering that

$$\tilde{\mathbf{p}}^{(0)} \prec_{ST} \mathbf{q} \prec_{ST} \tilde{\mathbf{p}}^{(1)}. \qquad (11.6)$$

Since both bounds have the same asymptotic approximations, these approximations hold also for the quasi-stationary distribution. □

In summary we note that the bodies and the near-end tails of all three distributions $\mathbf{p}^{(1)}$, $\mathbf{p}^{(0)}$ and **q** are approximately normal with the mean equal to $\bar{\mu}_1$ and with standard deviation $\bar{\sigma}_1$ in the parameter region where $R_0 > 1$ is fixed as $N \to \infty$. We note also that the important left tails of these distributions are all different. In particular, the tail probabilities $p_1^{(1)}$, $p_1^{(0)}$, and q_1 are simply related. We find

$$p_1^{(1)} \sim \frac{R_0}{\sqrt{N}} \varphi(\beta_1), \qquad R_0 > 1, \quad N \to \infty, \qquad (11.7)$$

$$q_1 \sim \frac{(R_0 - 1)^2}{R_0} \sqrt{N} \varphi(\beta_1), \quad R_0 > 1, \quad N \to \infty, \qquad (11.8)$$

$$p_1^{(0)} \sim (R_0 - 1) \sqrt{N} \varphi(\beta_1), \quad R_0 > 1, \quad N \to \infty. \qquad (11.9)$$

The three probabilities have been listed in increasing order. The smallest one is listed first, and the largest one is listed last. This order is consistent with the stochastic ordering given by (3.105), which implies that

$$\mathbf{p}^{(0)} \prec_{ST} \mathbf{q} \prec_{ST} \mathbf{p}^{(1)}. \qquad (11.10)$$

We note also that the ratios between successive probabilities in the above list are of different orders. Thus, $q_1/p_1^{(1)}$ is of the order of N, while $p_1^{(0)}/q_1$ is of the order of one.

In the parameter region where $R_0 > 1$, we find that $\bar{\mu}_1$ is of the order of N and that $\bar{\sigma}_1$ is of the order of $\sqrt{N}$. The coefficient of variation of the quasi-stationary distribution is therefore of the order of $1/\sqrt{N}$. This supports the arguments by May [41] that deterministic modelling is acceptable for sufficiently large populations if $R_0 > 1$. But the question of determining when N is sufficiently large remains. To answer it, we evaluate $\rho = (R_0 - 1)\sqrt{N}$. As a rule of thumb, we say that a community characterized by given values of R_0 and N belongs to the parameter region where R_0 is distinctly larger than one if ρ is larger than three. This parameter region is also the one where the deterministic model gives an acceptable approximation of the stochastic one.

We note also that early approximations of mean and standard deviation of the quasi-stationary distribution are included in Bartlett et al. [13]. These authors use the stationary distribution $\mathbf{p}^{(0)}$ as an approximation of the quasi-stationary distribution. The systematic study undertaken in the present monograph shows that this approach gives a good approximation of the quasi-stationary distribution both in its body and its near-end tails, but not in its left tail, when R_0 is distinctly larger than one. Therefore, the approximations of mean and standard deviation are acceptable. However, this approach cannot be expected to give satisfactory results concerning the quasi-stationary distribution in any of the remaining parameter regions, namely the transition region or the region where R_0 is distinctly below one.

Another method that has been used to find approximations of the first few moments of the quasi-stationary distribution is the so-called moment closure method. An early contribution dealing with the Verhulst model was given by Matis and Kiffe [40]. A later paper by Nåsell [49] also deals with the Verhulst model. It allows additional terms to be included in asymptotic approximations of the mean and the variance of the quasi-stationary distribution. Further extensions for the Verhulst model are provided by Singh and Hespana [68]. All of these papers deal only with the case when R_0 is distinctly above one. The recent paper on the SIS model by Clancy and Mendy [19] uses the moment closure method also in the case when R_0 is distinctly below one. They argue that the method is not well suited for investigations in the transition region.

11.2   Approximation of the Quasi-stationary Distribution q when R_0 is Distinctly Below One

We give here asymptotic approximations of the quasi-stationary distribution $\mathbf{q}$ in the parameter region where $R_0 < 1$. We include also approximations of the distribution function F, defined by

$$F(n) = \sum_{k=1}^{n} q_k, \quad n = 1, 2, \ldots, N. \tag{11.11}$$

The results are contained in the following theorem.

Theorem 11.2 *The quasi-stationary distribution* **q** *of the SIS model is approximated as follows when* $R_0 < 1$ *as* $N \to \infty$:

$$q_n \sim \frac{1 - R_0}{R_0} \frac{\varphi(y_2(n))}{\varphi(\beta_2)}, \quad R_0 < 1, \quad n = o(N), \quad 1 \le n, \quad N \to \infty, \tag{11.12}$$

$$q_n \sim (1 - R_0) R_0^{n-1}, \quad R_0 < 1, \quad n = o(\sqrt{N}), \quad 1 \le n, \quad N \to \infty, \tag{11.13}$$

$$q_1 \sim 1 - R_0, \quad R_0 < 1, \quad N \to \infty, \tag{11.14}$$

$$F(n) \sim 1 - \frac{\varphi(y_2(n))}{\varphi(\beta_2)}, \quad R_0 < 1, \quad n = o(N), \quad 1 \le n, \quad N \to \infty, \tag{11.15}$$

$$F(n) \sim 1 - R_0^n, \quad R_0 < 1, \quad n = o(\sqrt{N}), \quad 1 \le n, \quad N \to \infty. \tag{11.16}$$

Proof. These results follow readily from the corresponding results for $\Psi\left(\mathbf{p}^{(1)}\right)$, given in Sect. 10.2. The derivation is simple. We have shown in Sect. 10.2 that the two distributions $\mathbf{p}^{(1)}$ and $\tilde{\mathbf{p}}^{(1)} = \Psi\left(\mathbf{p}^{(1)}\right)$ have the same asymptotic approximations. This allows the conclusion that $\tilde{\mathbf{p}}^{(i,1)} = \Psi^i\left(\mathbf{p}^{(1)}\right)$ also has the same asymptotic approximation for $i = 1, 2, \ldots$. We now refer to the result by Ferrari et al. [27] that shows that the quasi-stationary distribution **q** is the limit of the distributions $\tilde{\mathbf{p}}^{(i,1)}$ as $i \to \infty$. It follows therefore that the quasi-stationary distribution **q** has the same asymptotic approximation, as given in (11.12)–(11.16). $\square$

A comparison with the results given by Nåsell [47] shows that the approximation of the left tail given by (11.12) is new, while its restriction to the left-most tail given in (11.13) was known in 2001.

We add here the comment that an important feature of the quasi-stationary distribution is that the distribution of the state variable approaches the quasi-stationary distribution as time goes on, and before extinction takes place. This property is of importance from a modelling standpoint. It is well satisfied in the parameter region treated in the previous section, with $R_0 > 1$, and reasonably well also in the transition region. However, this property holds poorly in the parameter region treated here, with $R_0 < 1$. Indeed, the time to extinction is here so short that one can expect extinction to take place before the distribution of the state variable has had time to reach the quasi-stationary distribution, if the initial distribution differs appreciably from the quasi-stationary distribution. It is important to recognize this aspect in the interpretation of our results concerning the quasi-stationary distribution.

11.3 Approximation of the Quasi-stationary Distribution q in the Transition Region

In this section we present an explicit approximation of the quasi-stationary distribution **q** in the transition region. We do not claim that the approximation is asymptotic. To describe the result, we define three functions H, $\tilde{\rho}$ and R by

$$H(y) = \frac{1}{y + 1/H(y)} \int_{-1/H(y)}^{y} H_1(t)\,dt, \quad y \in \mathfrak{R}, \tag{11.17}$$

$$\tilde{\rho}(y) = y + \frac{1}{H(y)}, \quad y \in \mathfrak{R}, \tag{11.18}$$

and

$$R(N,y) = 1 + \frac{\tilde{\rho}(y)}{\sqrt{N}} = 1 + \frac{y + 1/H(y)}{\sqrt{N}}. \tag{11.19}$$

A plot of H is included in Fig. 13.1, where also the related functions H_1 and H_0 are plotted.

For later use we note that $H(y)$ satisfies the following relation:

$$1 + yH(y) = \int_{-1/H(y)}^{y} H_1(t)\,dt. \tag{11.20}$$

We note that the functions H and H_i are similar, as are the functions $\tilde{\rho}$ and $\tilde{\rho}_i$, and the functions R and R_i, where functions with subscript i are defined in (10.89)–(10.91). Before stating the approximations that are established in this section, we prove that the sequence of functions H_i converges to H as $i \to \infty$. It follows from this that the sequence of functions $\tilde{\rho}_i$ converges to $\tilde{\rho}$, and that the sequence of functions R_i converges to R as $i \to \infty$.

To show that the sequence of functions H_i converges to H, we prove first that $H_1(y) > H_2(y) > H_3(y) > \cdots > 0$ holds for any real y, using induction. The inequalities

$$H_1(y) > H_2(y) > 0, \quad y \in \mathfrak{R}, \tag{11.21}$$

have been established in (10.64). We assume now that $H_1(y) > H_2(y) > \cdots > H_i(y) > 0$ and proceed to show that $H_i(y) > H_{i+1}(y) > 0$. Clearly, we have

$$y + \frac{1}{H_i(y)} > y + \frac{1}{H_1(y)} > 0. \tag{11.22}$$

By using the definition of H_i in (10.89) we get

$$H_{i+1}(y) = \frac{1}{y + 1/H_i(y)} \int_{-1/H_i(y)}^{y} H_1(t)\,dt, \quad y \in \mathfrak{R}. \tag{11.23}$$

The inequality in (11.22) shows that the lower limit of integration is smaller than the upper limit of integration. The integral is therefore positive, since the integrand $H_1(t)$ is. Furthermore, the factor in front of the integral is positive, again using (11.22). This leads to the inequality $H_{i+1}(y) > 0$. The upper bound $H_{i+1}(y) < H_i(y)$ follows from the fact that the function f defined by

$$f(x) = \frac{1}{y-x} \int_x^y H_1(t)\,dt \tag{11.24}$$

is an increasing function of x for $x < y$. To show this, we determine the derivative of f:

$$f'(x) = \frac{1}{(y-x)^2} \int_x^y H_1(t)\,dt - \frac{1}{y-x}H_1(x), \quad x < y. \tag{11.25}$$

The integrand is larger than $H_1(x)$, since the function H_1 is increasing. Hence the integral is larger than $(y-x)H_1(x)$. We conclude therefore that $f'(x) > 0$. This completes the proof that the sequence $H_i(y)$ is decreasing in i for each real y. Since it is also bounded below by zero, we conclude that the limit as $i \to \infty$, $H(y)$, exists.

The approximations in this section are given by the following theorem:

Theorem 11.3 *The quasi-stationary distribution* **q** *of the SIS model is approximated as follows in the transition region:*

$$q_n \approx \frac{1}{\varphi(\rho)\tilde{\rho}(\rho)H(\rho)} \frac{1}{n}\left(1 - \frac{1}{[R(N,\rho)]^n}\right)\varphi(y_3(n)),$$

$$\rho = O(1), \quad n = o(N), \quad 1 \le n, \quad N \to \infty, \quad (11.26)$$

$$q_n \approx \frac{1}{H(\rho)\sqrt{N}}, \quad \rho = O(1), \quad n = o(\sqrt{N}), \quad 1 \le n, \quad N \to \infty, \tag{11.27}$$

$$q_1 \approx \frac{1}{H(\rho)\sqrt{N}}, \quad \rho = O(1), \quad N \to \infty. \tag{11.28}$$

Proof. To derive the approximation of the quasi-stationary distribution **q** given in (11.26), we use the fact established by Ferrari et al. [27] that the sequence $\Psi^i\left(\mathbf{p}^{(1)}\right)$ converges to the quasi-stationary distribution **q** as $i \to \infty$. The corresponding sequence of approximations of $\Psi^i\left(\mathbf{p}^{(1)}\right)$ given in (10.96) will therefore converge to an approximation of the quasi-stationary distribution **q**. The limit of this sequence of approximations is given by (11.26). This concludes the derivation of the approximation (11.26) of the quasi-stationary distribution **q** in the transition region.

We show next that the approximation in (11.27) follows from the approximation in (11.26). We get

$$\log[R(N,\rho)]^n = n\log\left(1 + \frac{\tilde{\rho}(\rho)}{\sqrt{N}}\right) \sim \tilde{\rho}(\rho)\frac{n}{\sqrt{N}}, \quad \rho = O(1), \quad N \to \infty. \tag{11.29}$$

This approximation approaches zero as $N \to \infty$ if $n = o(\sqrt{N})$. Hence,

$$[R(N,\rho)]^n \sim 1 + \tilde{\rho}(\rho)\frac{n}{\sqrt{N}}, \quad n = o(\sqrt{N}), \quad 1 \leq n, \quad N \to \infty. \tag{11.30}$$

This gives

$$1 - \frac{1}{[R(N,\rho)]^n} = \frac{[R(N,\rho)]^n - 1}{[R(N,\rho)]^n} \sim \tilde{\rho}(\rho)\frac{n}{\sqrt{N}}, \quad n = o(\sqrt{N}), \quad 1 \leq n, \quad n \to \infty.$$

$$\tag{11.31}$$

The result in (11.27) follows by inserting this approximation of $1 - 1/[R(N,\rho)]^n$, together with the approximation $\varphi(y_3(n))/\varphi(\rho) \sim 1$ in (7.32), into (11.26).   □

Chapter 12
Approximation of the Time to Extinction for the SIS Model

We noted in Sect. 3.4, (3.42), that the time to extinction τ_Q from quasi-stationarity for a birth–death process with finite state space and with an absorbing state at the origin has an exponential distribution with expectation equal to $E\tau_Q = 1/(\mu_1 q_1)$. We noted also, in (3.44), that the expected time to extinction from the state 1 equals $E\tau_1 = 1/(\mu_1 p_1^{(0)})$. Approximations of these expectations for the SIS model are therefore easily established, since we have derived approximations of the two probabilities q_1 and $p_1^{(0)}$ in Chaps. 11 and 9, respectively, and since $\mu_1 = \mu$.

The expected time to extinction from the state n can be expressed as follows, using (3.45):

$$E\tau_n = E\tau_1 S_n^{(0)}, \quad n = 1, 2, \dots, N, \tag{12.1}$$

where $S_n^{(0)}$ is defined in (10.3).

We present approximations of $E\tau_Q$ and of $E\tau_1$ separately in each of the three parameter regions, while we give approximations of $E\tau_n$ for $n > 1$ only in the parameter region where R_0 is distinctly above one. We leave it as an open problem to derive approximations of $E\tau_n$ for $n > 1$ in the transition region and in the region where R_0 is distinctly below one.

12.1 Approximation of the Time to Extinction when R_0 is Distinctly Above One

This section gives approximations of the expected time to extinction from quasi-stationarity, $E\tau_Q$, and of the expected time to extinction from the state n, $E\tau_n$, in the parameter region where R_0 is distinctly larger than one. Approximations of the latter quantity are limited to n-values that are of smaller order than N. We give also a simplified expressions for $E\tau_n$ when n is of smaller order than $\sqrt{N}$, and in the special case when $n = 1$. The results are given in the following theorem.

I. Nåsell, *Extinction and Quasi-stationarity in the Stochastic Logistic SIS Model*, 149
Lecture Notes in Mathematics 2022, DOI 10.1007/978-3-642-20530-9_12,
© Springer-Verlag Berlin Heidelberg 2011

Theorem 12.1 *The expected time to extinction for the SIS model is approximated as follows when $R_0 > 1$ as $N \to \infty$ in the cases when initially quasi-stationarity holds $(E\tau_Q)$, and when initially there are n infected individuals $(E\tau_n)$:*

$$E\tau_Q = \frac{1}{\mu q_1} \sim \frac{\sqrt{2\pi}}{\mu} \frac{R_0}{(R_0-1)^2} \frac{\exp(\gamma_1 N)}{\sqrt{N}}, \quad R_0 > 1, \quad N \to \infty, \quad (12.2)$$

$$E\tau_n \sim \frac{\sqrt{2\pi}}{\mu} \frac{R_0}{(R_0-1)^2} \frac{\exp(\gamma_1 N)}{\sqrt{N}} \left(1 - \frac{\varphi(\beta_2)}{\varphi(y_2(n))}\right),$$

$$R_0 > 1, \quad n = o(N), \quad 1 \le n, \quad N \to \infty, \quad (12.3)$$

$$E\tau_n \sim \frac{\sqrt{2\pi}}{\mu} \frac{R_0}{(R_0-1)^2} \frac{\exp(\gamma_1 N)}{\sqrt{N}} \left(1 - \frac{1}{R_0^n}\right),$$

$$R_0 > 1, \quad n = o(\sqrt{N}), \quad 1 \le n, \quad N \to \infty, \quad (12.4)$$

$$E\tau_1 = \frac{1}{\mu p_1^{(0)}} \sim \frac{\sqrt{2\pi}}{\mu} \frac{1}{R_0-1} \frac{\exp(\gamma_1 N)}{\sqrt{N}}, \quad R_0 > 1, \quad N \to \infty. \quad (12.5)$$

Proof. The result in (12.2) follows by inserting the approximation (11.5) for q_1, valid in the parameter region where $R_0 > 1$, into the expression $E\tau_Q = 1/(\mu q_1)$, and also noting the relation $\varphi(\beta_1) = \exp(-\gamma_1 N)/\sqrt{2\pi}$ in (6.50).

The approximation of $E\tau_1$ in (12.5) follows by inserting the approximation (9.5) of $p_1^{(0)}$ into the expression $E\tau_1 = 1/(\mu p_1^{(0)})$, again using the relation (6.50).

The approximation (12.3) of $E\tau_n$ for $n = o(N)$ follows by inserting the approximation of $E\tau_1$ in (12.5) and the approximation of $S_n^{(0)}$ in (10.16) into the expression $E\tau_n = E\tau_1 S_n^{(0)}$ in (12.1).

Finally, the approximation (12.4) of $E\tau_n$ for $n = o(\sqrt{N})$ follows from the approximation in (12.3) and the approximation $\varphi(y_2(n))/\varphi(\beta_2) \sim R_0^n$ in (6.45), valid for $n = o(\sqrt{N})$.   □

The expression in (12.2) shows that the approximation of the expected time to extinction from quasi-stationarity grows exponentially with N when $R_0 > 1$, since $\gamma_1 > 0$ for $R_0 > 1$.

It should be noted that the time to extinction from state n, τ_n, is a random variable whose location is poorly measured by its expectation $E\tau_n$ for small values of n when $R_0 > 1$. The reason for this is that its distribution is bimodal. In fact, τ_n can for small values of n be seen as a mixture of two random variables with widely different expectations. We verify this with $n = 1$. The process will then either reach the absorbing state 0 very quickly, with probability $1/R_0$, or it will with the complementary probability $1 - 1/R_0$ reach the set of states where the quasi-stationary distribution describes its behavior, and where the expected time to extinction is equal to τ_Q. By ignoring the time to extinction when the process takes the first of these paths, we are led to approximate $E\tau_1$ by the probability $1 - 1/R_0$ multiplied by the expected time to extinction from quasi-stationarity $E\tau_Q$. It is easy to see that the ratio of the two approximations given by (12.2) and (12.5) is $E\tau_1/E\tau_Q \sim 1 - 1/R_0$, which is the same as above.

The approximation (12.2) is not new; it was essentially given both by Barbour [8] and by Andersson and Djehiche [4]. Indeed, Barbour's result shows that if T_N denotes the time to extinction, then

$$\lim_{N \to \infty} P[k_N T_N \geq x] = \exp(-x), \tag{12.6}$$

where

$$k_N = \frac{(\bar{\alpha}_1 - \bar{\alpha}_2)^2}{\bar{\gamma}_1 + \bar{\gamma}_2} \sqrt{\left(\frac{\bar{\gamma}_1}{\bar{\alpha}_1} + \frac{\bar{\gamma}_2}{\bar{\alpha}_2}\right) \frac{N}{2\pi}}$$

$$\times \left(\frac{\bar{\alpha}_1 \bar{\gamma}_2 + \bar{\alpha}_2 \bar{\gamma}_1}{\bar{\alpha}_1 (\bar{\gamma}_1 + \bar{\gamma}_2)}\right)^{N\bar{\alpha}_1/\bar{\gamma}_1} \left(\frac{\bar{\alpha}_1 \bar{\gamma}_2 + \bar{\alpha}_2 \bar{\gamma}_1}{\bar{\alpha}_2 (\bar{\gamma}_1 + \bar{\gamma}_2)}\right)^{N\bar{\alpha}_2/\bar{\gamma}_2}. \tag{12.7}$$

Here, the parameters $\bar{\alpha}_1$, $\bar{\gamma}_1$, $\bar{\alpha}_2$, and $\bar{\gamma}_2$ are related to our transition rates λ_n and μ_n as follows:

$$\lambda_n = \bar{\alpha}_1 n - \bar{\gamma}_1 \frac{n^2}{N}, \qquad \mu_n = \bar{\alpha}_2 n + \bar{\gamma}_2 \frac{n^2}{N}. \tag{12.8}$$

Barbour's parameters can therefore be expressed in ours by the relations

$$\bar{\alpha}_1 = \lambda, \quad \bar{\gamma}_1 = \lambda, \quad \bar{\alpha}_2 = \mu, \quad \bar{\gamma}_2 = 0. \tag{12.9}$$

This means that Barbour deals with the more general Verhulst model, where both infection rate λ_n and recovery rate μ_n show density dependence.

The first three factors that appear in the right-hand side of (12.7) are easy to evaluate for these values of Barbour's parameters. For the fourth factor, we need to exercise some care to find its value as $\bar{\gamma}_2 \to 0$. Thus, direct insertion of the above parameter values, except $\bar{\gamma}_2$, shows that the fourth factor can be written

$$\left[\frac{1 + \bar{\gamma}_2/\mu}{1 + \bar{\gamma}_2/\lambda}\right]^{N\mu/\bar{\gamma}_2}. \tag{12.10}$$

The limit of this factor as $\bar{\gamma}_2 \to 0$ is easily shown to be equal to

$$\exp\left(N\left(1 - \frac{\mu}{\lambda}\right)\right), \tag{12.11}$$

for example by showing that its logarithm approaches $N(1 - \mu/\lambda)$ as $\bar{\gamma}_2 \to 0$.

The expression for k_N can therefore be written

$$k_N = \frac{(\lambda - \mu)^2}{\lambda} \sqrt{\frac{N}{2\pi}} \left(\frac{\mu}{\lambda}\right)^N \exp\left(N\left(1 - \frac{\mu}{\lambda}\right)\right)$$

$$= \mu \frac{(R_0 - 1)^2}{R_0} \sqrt{\frac{N}{2\pi}} \left(\frac{1}{R_0}\right)^N \exp\left(N\left(1 - \frac{1}{R_0}\right)\right) \qquad (12.12)$$

Barbour's result for finite N and $R_0 > 1$ is that the distribution of the time to extinction T_N from an arbitrary initial distribution is approximately exponential with the expected value $1/k_N$. This can be compared with our result that if the initial distribution equals the quasi-stationary distribution, then the distribution of τ_Q is exactly exponential with expectation approximated by (12.2). It is straightforward to use the above expression for k_N to show that Barbour's expression for $1/k_N$ is equal to our approximation (12.2) for $E\tau_Q$. Our parametrization has the advantage of giving a possibly better heuristic understanding for the N-dependence of $E\tau_Q$.

Andersson and Djehiche [4] study the extinction time τ_n. They are concerned with its distribution, while our results above are restricted to deal with expectations. They show that the distribution of τ_n approaches an exponential distribution with the expectation given by (12.2) if the initial number of infected individuals is of the order of N. This result is quite different from our result for τ_n, since we do not assume that n grows with N. Andersson and Djehiche also study τ_n for $n = O(1)$, and show that this random variable is a mix of one that approaches the extinction time for a birth–death process with linear transition rates, and one that approaches the exponential distribution as above. These results are described by Andersson and Britton [3].

Newman et al. [51] also analyze the extinction time τ_1. They use Laplace transforms to derive the approximation (12.5) for its expectation. They consider also its coefficient of variation, which shows unexpected behavior as a function of R_0. Further insight into this behavior might benefit from an approach similar to that used by Andersson and Djehiche that treats the extinction time τ_1 as a mix of two variables with widely different expectations.

12.2 Approximation of the Time to Extinction when R_0 is Distinctly Below One

This section gives approximations of the expected times to extinction from quasi-stationarity and from the state one in the parameter region where $R_0 < 1$. The results are as follows:

Theorem 12.2 *The expected time to extinction for the SIS model is approximated as follows when $R_0 < 1$ as $N \to \infty$ in the two cases when initially quasi-stationarity holds ($E\tau_Q$), and when initially there is one infected individual ($E\tau_1$):*

$$E\tau_Q = \frac{1}{\mu q_1} \sim \frac{1}{\mu}\frac{1}{1-R_0}, \quad R_0 < 1, \quad N \to \infty. \tag{12.13}$$

$$E\tau_1 = \frac{1}{\mu p_1^{(0)}} \sim \frac{1}{\mu}\frac{\log(1/(1-R_0))}{R_0}, \quad R_0 < 1, \quad N \to \infty. \tag{12.14}$$

Proof. The approximation (12.13) of the expected time to extinction from quasi-stationarity in the parameter region where $R_0 < 1$ as $N \to \infty$ is found by inserting the approximation (11.14) of the probability q_1 into the expression $E\tau_Q = 1/(\mu q_1)$.

Similarly, the approximation of the expected time to extinction from the state one given in (12.14) is found by inserting the approximation (9.12) of the probability $p_1^{(0)}$ into the expression $E\tau_1 = 1/(\mu p_1^{(0)})$.   □

We note that these approximations of expectations of time to extinction are independent of N when $R_0 < 1$, while the results in the previous section show that the corresponding approximations grow exponentially with N when $R_0 > 1$.

Andersson and Djehiche [4] also study the random variable τ_n for $R_0 \le 1$. They show that if $n = O(N)$, then a linear combination $a\tau_n + b$ can be found that approaches an extreme value distribution. On the other hand, if $n = O(1)$, then τ_n approaches a random variable which is the extinction time for a birth–death process with linear transition rates. The work of Andersson and Djehiche has not been extended to deal with the transition region.

12.3   Approximation of the Time to Extinction in the Transition Region

The expected time to extinction from quasi-stationarity $E\tau_Q$ and from the state one $E\tau_1$ are approximated as follows in the transition region:

Theorem 12.3 *The expected time to extinction for the SIS model is approximated as follows in the transition region in the two cases when initially quasi-stationarity holds ($E\tau_Q$), and when initially there is one infected individual ($E\tau_1$):*

$$E\tau_Q = \frac{1}{\mu_1 q_1} \approx \frac{1}{\mu}\sqrt{N}H(\rho), \quad \rho = O(1), \quad N \to \infty, \tag{12.15}$$

$$E\tau_1 = \frac{1}{\mu_1 p_1^{(0)}} \approx \frac{1}{\mu}\left(\frac{1}{2}\log N + H_0(\rho)\right), \quad \rho = O(1), \quad N \to \infty. \tag{12.16}$$

Proof. The approximation (12.15) of $E\tau_Q$ in the transition region follows by inserting the approximation (11.27) of the probability q_1 into the expression $E\tau_Q = 1/(\mu_1 q_1)$ in (3.42), and noting that $\mu_1 = \mu$ for the SIS model.

The approximation (12.16) of $E\tau_1$ in the transition region follows by inserting the approximation (9.15) of the probability $p_1^{(0)}$ into the expression $E\tau_1 = 1/(\mu_1 p_1^{(0)})$ in (3.44), and also noting, as above, that $\mu_1 = \mu$ for the SIS model.   □

We note that the expected time to extinction from the state n, $E\tau_n$, grows quite strongly with n in the transition region. The reason is based on our results that $E\tau_Q$ is of the order $\sqrt{N}$, while $E\tau_1$ is of the order $\log N$. Hence the relation $E\tau_Q = \sum_{n=1}^{N} q_n E\tau_n$ implies that there are n-values for which $E\tau_n$ is at least of the order $\sqrt{N}$. This in turn implies that the ratio $E\tau_n/E\tau_1$ is at least of the order $\sqrt{N}/\log N$ for these values of n.

The ratio $E\tau_Q/E\tau_1$ is seen to be larger than one in all three parameter regions, as can be expected heuristically. Furthermore, it is bounded as $N \to \infty$ in the two parameter regions where R_0 is distinctly larger than one and where R_0 is distinctly smaller than one, while it grows with N in the transition region. Thus, this ratio has a maximum as a function of ρ for some finite value of ρ for fixed N. This behavior is consistent with the finding that the ratio is an increasing function of R_0 for $R_0 < 1$, and a decreasing function of R_0 for $R_0 > 1$, again for fixed N.

Newman et al. [51] also study the extinction time τ_1 in the transition region. They use what they call a crude saddle-point calculation to find an approximation of $E\tau_1$ in the transition region. Expressed with our notation, their result can be written as follows:

$$E\tau_1 \approx \frac{1}{\mu}\left(\frac{1}{2}\log N + \frac{1}{\rho\varphi(\rho)} + \frac{\exp(\rho - 1/2)}{\rho}\right), \quad \rho = O(1), \quad \rho > 1, \quad N \to \infty.$$

(12.17)

A numerical evaluation shows that this is reasonably close to our approximation in (12.16). It has, however, the disadvantage that it covers only part of the transition region.

The so-called decay parameter, which equals the inverse of the expected time to extinction from quasi-stationarity, is studied by Sirl et al. [69]. They establish both lower and upper bounds of the decay parameter. They use numerical evaluations to show that these bounds are for some parameter combinations considerably closer to the actual value of the decay parameter than the approximations that we have presented in this chapter. The bounds given by Sirl et al. are based on results of Chen [17].

Chapter 13
Uniform Approximations for the SIS Model

Our main concern in this monograph is with the quasi-stationary distribution and the time to extinction for the stochastic logistic SIS model. Explicit expressions are not available for these quantities. Furthermore, they behave in qualitatively different ways in three different parameter regions. Because of this, we have derived approximations for each quantity, and by necessity we have used different arguments to derive such approximations in each of the three parameter regions. The need to give three separate formulas to describe the approximation of any one quantity is somewhat cumbersome and unelegant. To overcome this weakness, we give here uniform approximations of the various quantities, valid across all three of the parameter regions. We include also uniform approximations of the stationary distributions $\mathbf{p}^{(1)}$ and $\mathbf{p}^{(0)}$ of the two auxiliary processes $\{I^{(1)}(t)\}$ and $\{I^{(0)}(t)\}$.

The quasi-stationary distribution $\mathbf{q}$, as well as the stationary distributions $\mathbf{p}^{(1)}$ and $\mathbf{p}^{(0)}$, are determined by the two essential parameters N and R_0, and these are also the parameters that are used for describing the approximations of these quantities in the two parameter regions where R_0 is distinctly different from one. However, a different parametrization is used in the transition region where R_0 is close to one, namely N and ρ, where $\rho = (R_0 - 1)\sqrt{N}$. The transition region is characterized by finite values of ρ, which means that R_0 approaches the value one in the transition region as $N \to \infty$.

For the purposes of deriving uniform results, we use the same parametrization in all three parameter regions, given by N and ρ. Finite values of ρ will then describe the transition region, while results in the two parameter regions where R_0 is distinctly different from one are found by letting ρ approach $\pm\infty$ for fixed N.

The starting point for the derivation of a uniform approximation is the approximation that we have derived in the transition region. The first question to ask is whether the limits of this result when $\rho \to \pm\infty$ give the results derived in the two parameter regions where R_0 is distinctly different from one. It turns out that this is nowhere the case. We proceed therefore to introduce what may be called innocent changes in the approximation derived in the transition region. This term describes changes that do not affect the asymptotic approximation as $N \to \infty$ in

I. Nåsell, *Extinction and Quasi-stationarity in the Stochastic Logistic SIS Model*, 155
Lecture Notes in Mathematics 2022, DOI 10.1007/978-3-642-20530-9_13,
© Springer-Verlag Berlin Heidelberg 2011

the same region. There are several examples of such changes. Since $R_0 \sim 1$ in the transition region, we can multiply or divide the original approximation by e.g. R_0 or $\sqrt{R_0}$. Furthermore, there are several quantities that appear in the original approximation that can be exchanged for each other. The four quantities $\beta_0 = \rho$, β_1, β_2, and β_3 are exchangeable in this way, since each of the latter three can be shown to be asymptotic to ρ in the transition region as $N \to \infty$. (The two-term asymptotic approximations of them are given in Appendix A.) As a further example of an innocent change we mention the possibility of multiplying the originally derived approximation by the ratio of any two of these quantities. Another set of exchangeable quantities is given by $y_1(n)$, $y_2(n)$, and $y_3(n)$.

The derivation of a uniform approximation starts from the originally derived approximation in the transition region. Two different versions of this approximation are then derived by subjecting it to innocent changes. The two versions are chosen such that the extensions to the two parameter regions where R_0 is distinctly different from one agree asymptotically with the approximations that have been derived previously in the corresponding parameter region.

The extension of approximations from the transition region to the two regions where R_0 is distinctly different from one requires knowledge about the asymptotic behaviors of the functions H_1, H_0, and H as their arguments approach $\pm\infty$. We also need information about the behavior of $R(N,y) = 1 + (y + 1/H(y))/\sqrt{N}$ as $y \to \pm\infty$. Results of this nature are given in Sect. 13.1.

In line with what has been seen in the non-uniform results presented earlier on in this monograph, we shall give approximations of the three distributions $\mathbf{p}^{(1)}$, $\mathbf{p}^{(0)}$, and $\mathbf{q}$ both in their bodies and in their left tails. All such results will hold under certain restrictions on the state variable values n for which the approximations are valid. We shall encounter one problem in the presentation of uniform results of this type. The problem occurs in the discussion of results that are valid in the bodies of the distributions in question. It is caused by the fact that only positive values are possible for all three distributions. This means that the body of the distribution will actually lie in the left tail of the distribution in the two parameter regions where $R_0 < 1$ and where $\rho = O(1)$, while there is a large separation between body and left tail in the case when $R_0 > 1$. Because of this, the description of the restrictions on n will not itself be uniform in these cases.

On the other hand, the left tail of any one of the three distributions is described by the condition that $n = o(N)$, and the left-most tail by the condition $n = o(\sqrt{N})$. In both cases we find that the conditions are independent of the parameter region, and therefore can be regarded as uniform over the parameter regions.

The approximation in the body of any distribution will also hold in its near-end tails. The approximations of the left-most tails are somewhat simpler than the approximations of the left tails. We shall also give uniform approximations of the probabilities $p_1^{(1)}$, $p_1^{(0)}$, and q_1. The last two of these probabilities are important for determining the expected time to extinction from the state one and from the quasi-stationary distribution, respectively, as shown in Sect. 3.4. Uniform approximations of the stationary distributions $\mathbf{p}^{(1)}$ and $\mathbf{p}^{(0)}$ are given in Sects. 13.2 and 13.3,

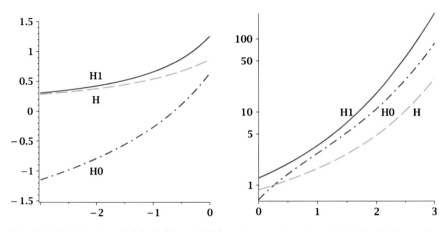

Fig. 13.1 The functions $H_1(y)$, $H_0(y)$, and $H(y)$ are shown as functions of y. In the *left* figure, the argument y is negative, and the vertical scale is linear. In the *right* figure, the argument y is positive, and the vertical scale is logarithmic

while uniform approximations of the quasi-stationary distribution $\mathbf{q}$ are found in Sect. 13.4. Finally, uniform approximations of the expected time to extinction from quasi-stationarity, $E\tau_Q$, and of the expected time to extinction from the state one, $E\tau_1$, are given in Sect. 13.5. The last two sections summarize the most important results derived in the monograph.

The behaviors of the functions H_1, H_0, and H are shown by the plots in Fig. 13.1. The notation used for these three functions is chosen to remind the reader which of the three important distributions $\mathbf{p}^{(1)}$, $\mathbf{p}^{(0)}$, and $\mathbf{q}$ that the function is used to describe. Thus, H_1 appears in approximations of the stationary distribution $\mathbf{p}^{(1)}$, H_0 in approximations of the stationary distribution $\mathbf{p}^{(0)}$, and H in approximations of the quasi-stationary distribution $\mathbf{q}$.

13.1   Asymptotic Approximations of $H_1, H_0, H,$ and R

In this section we give asymptotic approximations of H_1, H_0, H, and R as $\rho \to \pm\infty$. The asymptotic behavior of the function H_1 as its argument approaches $\pm\infty$ was studied in Sect. 5.3, and given by Corollary 5.3. We repeat it here for ease of reference, and for comparison with the asymptotic approximations of the functions H_0 and H:

$$H_1(y) \sim \begin{cases} \dfrac{1}{\varphi(y)}, & y \to +\infty, \\[2ex] -\dfrac{1}{y}, & y \to -\infty. \end{cases} \tag{13.1}$$

The asymptotic behaviors of the functions H_0, H, and R are derived below. The results are summarized as follows:

$$H_0(y) \sim \begin{cases} \dfrac{1}{y\varphi(y)}, & y \to +\infty, \\[2mm] -\log|y|, & y \to -\infty, \end{cases} \tag{13.2}$$

and

$$H(y) \sim \begin{cases} \dfrac{1}{y^2\varphi(y)}, & y \to +\infty, \\[2mm] -\dfrac{1}{y} + \dfrac{2}{y^3}, & y \to -\infty. \end{cases} \tag{13.3}$$

In distinction to R_0, the function R defined in (11.19) is larger than one for all values of ρ. It behaves as follows:

$$R(N,y) \sim \begin{cases} 1 + \dfrac{y}{\sqrt{N}}, & y \to +\infty, \\[2mm] 1 - \dfrac{2}{y\sqrt{N}}, & y \to -\infty. \end{cases} \tag{13.4}$$

We proceed to derive these results. The asymptotic behavior of $H_0(y)$ as $y \to -\infty$ follows from the definition of the function H_0 in (8.10). To deal with the remaining statements above, we note that both H_0 and H are defined by integrals where the integrand equals $H_1(t)$. To deal with these integrals, we define an auxiliary function $\widetilde{H}_1$ as follows:

$$\widetilde{H}_1(y) = \int_0^y H_1(t)\,dt. \tag{13.5}$$

We show that

$$\widetilde{H}_1(y) \sim \frac{1}{y\varphi(y)}, \quad y \to \infty, \tag{13.6}$$

and that

$$|\widetilde{H}_1(y)| < \sqrt{\frac{\pi}{2}}|y|, \quad y < 0. \tag{13.7}$$

The integrand in the integral defining $\widetilde{H}_1(y)$ is written as follows:

$$H_1(t) = \frac{\Phi(t)}{\varphi(t)} = \frac{1 - \Phi(-t)}{\varphi(t)} = \varphi^*(t) - H_1(-t), \tag{13.8}$$

using the reciprocal φ^* of the normal density function φ. We get therefore

$$\widetilde{H}_1(y) = \Phi^*(y) - \int_0^y H_1(-t)\,dt, \tag{13.9}$$

where the integral Φ^* of the reciprocal of φ is defined by (5.10). The integral in this expression for $\widetilde{H}_1(y)$ is positive for $y > 0$. Furthermore, the integrand is smaller than $H_1(0) = \sqrt{2\pi}/2$, since H_1 is an increasing function. It follows that the integral obeys the following inequality:

$$\int_0^y H_1(-t)\,dt < \sqrt{\frac{\pi}{2}}\,y, \quad y > 0. \tag{13.10}$$

This term is of smaller magnitude than the first term in (13.9) for large y. We get therefore

$$\widetilde{H}_1(y) \sim \Phi^*(y) \sim \frac{1}{y\varphi(y)}, \quad y \to \infty, \tag{13.11}$$

using the asymptotic approximation of $H_1^*(y)$ given in Corollary 5.3. Thus, the approximation of $\widetilde{H}_1(y)$ given in (13.6) for large y-values has been derived.

With $y < 0$ we note first from (13.5) that $\widetilde{H}_1(y) < 0$. Furthermore, the integrand $H_1(t)$ is an increasing function of t. Thus, it is bounded by $H_1(0) = \sqrt{2\pi}/2$. The absolute value of the integral is therefore bounded by $H_1(0)|y|$. This is equivalent to the bound given in (13.7).

We now consider $H_0(y)$ for large values of y. From its definition in (8.10) we find that $H_0(y)$ can be expressed as follows:

$$H_0(y) = H_a(\rho_b) + \int_{\rho_b}^0 H_1(t)\,dt + \widetilde{H}_1(y), \quad y > 0, \tag{13.12}$$

where ρ_b is negative. Here, the first term $H_a(\rho_b)$ is constant, independent of y. It is of the order of one as $y \to \infty$. The second term is also of the order of one, as found by using the bound (13.7). We conclude therefore that

$$H_0(y) \sim \widetilde{H}_1(y) \sim \frac{1}{y\varphi(y)}, \quad y \to \infty, \tag{13.13}$$

using the asymptotic approximation of $\widetilde{H}_1(y)$ given in (13.6). Thus, both the approximations of $H_0(y)$ given in (13.2) have been derived.

We proceed to derive asymptotic approximations of $H(y)$ as $y \to \pm\infty$. We consider first the case $y \to \infty$. We note from (11.20) that $H(y)$ satisfies the relation

$$1 + yH(y) = \int_{-1/H(y)}^y H_1(t)\,dt, \tag{13.14}$$

and that therefore $H(y)$ can be expressed as follows:

$$H(y) = \frac{\widetilde{H}_1(y) - \widetilde{H}_1(-1/H(y)) - 1}{y}. \tag{13.15}$$

The first term in the numerator of this expression is exponentially large as $y \to \infty$, as seen from (13.6). The second term is of smaller magnitude, as seen from its bound in (13.7), and the third term is of still smaller magnitude. We conclude therefore that

$$H(y) \sim \frac{\widetilde{H}_1(y)}{y} \sim \frac{1}{y^2 \varphi(y)}, \qquad y \to \infty, \tag{13.16}$$

using the asymptotic approximation of $\widetilde{H}_1(y)$ as $y \to \infty$ given in (13.6). This establishes the asymptotic approximation of $H(y)$ as $y \to \infty$ given in (13.3).

Now consider $H(y)$ as $y \to -\infty$. We shall determine the two terms of the asymptotic approximation of $H(y)$ as $y \to -\infty$ given in (13.3). The basis for the derivation is as above the relation (13.14).

We insert two terms in the asymptotic approximation of the integrand $H_1(t)$ into the integrand in the integral of the right-hand side of (13.14). The two terms to use are found from (5.36) to be as follows:

$$H_1(y) \sim A_2(y) = -\frac{1}{y} + \frac{1}{y^3}, \qquad y \to -\infty. \tag{13.17}$$

Thus, we get

$$1 + yH(y) \sim \int_{-1/H(y)}^{y} \left[-\frac{1}{t} + \frac{1}{t^3} \right] dt = -\log(-yH(y)) - \frac{1}{2y^2} + \frac{1}{2}H^2(y), \quad y \to -\infty. \tag{13.18}$$

It does not appear possible to solve this equation explicitly for $H(y)$. Therefore, we look for an asymptotic approximation of $H(y)$ of the form

$$H(y) \sim -\frac{A}{y} + \frac{B}{y^3}, \qquad y \to -\infty. \tag{13.19}$$

Insertion of this assumed form into the relation above gives

$$1 - A + \frac{B}{y^2} \sim -\log A + \frac{B}{Ay^2} + \frac{1}{2}\frac{B^2}{A^2 y^4} + \frac{1}{2}\frac{A^2 - 1}{y^2} - \frac{AB}{y^4}, \qquad y \to -\infty. \tag{13.20}$$

We determine A and B so that the coefficients of the various powers of $1/y^2$ are equal to zero. This leads to the following three equations:

$$1 - A = -\log A, \tag{13.21}$$

$$B = \frac{1}{2}(A^2 - 1) + \frac{B}{A}, \tag{13.22}$$

$$AB = \frac{1}{2}\frac{B^2}{A^2}. \tag{13.23}$$

The first equation has the solution $A = 1$, while the second equation with $A = 1$ gives no condition on B, and the third equation, with $A = 1$, leads to $B = 2$. This establishes the asymptotic approximation of $H(y)$ given in (13.3) as $y \to -\infty$. Additional terms in this asymptotic approximation can be found by using additional terms in the approximation of the integrand $H_1(t)$.

The approximation of $R(N,y)$ as $y \to \infty$ in (13.4) follows from the approximation of $H(y)$ as $y \to \infty$ in (13.3) by insertion into the definition of $R(N,y)$ in (11.19).

To derive the approximation of $R(N,y)$ as $y \to -\infty$, we insert the approximation of $H(y)$ as $y \to -\infty$ in (13.3) into the definition of $R(N,y)$ in (11.19). We get

$$R(N,y) \sim 1 + \left(y + \frac{1}{-1/y + 2/y^3}\right)\frac{1}{\sqrt{N}} = 1 + \left(y - \frac{y}{1 - 2/y^2}\right)\frac{1}{\sqrt{N}}$$

$$\sim 1 + \frac{y - y(1 + 2/y^2)}{\sqrt{N}} = 1 - \frac{2}{y\sqrt{N}}, \quad y \to -\infty, \tag{13.24}$$

as claimed in (13.4).

13.2   Uniform Approximation of the Stationary Distribution p$^{(1)}$

In this section we give several uniform approximations of the stationary distribution $\mathbf{p}^{(1)}$. The first approximation given below deals with the body and near-end tails of the distribution. As has been mentioned above, the conditions on n in this case are not uniform over the parameter regions. This makes the statement of the conditions on n somewhat awkward. The second approximation in the theorem is concerned with the left tail, and the third one with the left-most tail. Finally, the fourth statement gives a uniform approximation of the probability $p_1^{(1)}$. The results are summarized in the following theorem. We refer to Appendix A for the notation that is used here.

Theorem 13.1 *The stationary distribution* $\mathbf{p}^{(1)}$ *of the SIS model has the following uniform approximation as* $N \to \infty$:

$$p_n^{(1)} \sim \frac{1}{H_1(\beta_{13})} \frac{1}{\bar{\sigma}_{12}} \frac{\varphi(y_{12}(n))}{\varphi(\beta_{12})},$$

$$y_1(n) = o(\sqrt{N}) \text{ when } R_0 > 1 \text{ as } N \to \infty,$$

$$\text{or } n \geq 1 \text{ and } n = o(N) \text{ when } \rho = O(1) \text{ as } N \to \infty,$$

$$\text{or } n \geq 1 \text{ and } n = o(N) \text{ when } R_0 < 1 \text{ as } N \to \infty, \tag{13.25}$$

$$p_n^{(1)} \sim \frac{1}{H_1(\beta_{13})} \frac{1}{\bar{\sigma}_2} \frac{\varphi(y_2(n))}{\varphi(\beta_2)}, \quad n = o(N), \quad 1 \leq n, \quad N \to \infty, \tag{13.26}$$

$$p_n^{(1)} \sim \frac{1}{H_1(\beta_{13})} \frac{R_0^n}{\bar{\sigma}_2}, \quad n = o(\sqrt{N}), \quad 1 \leq n, \quad N \to \infty, \tag{13.27}$$

$$p_1^{(1)} \sim \frac{1}{H_1(\beta_{13})} \frac{R_0}{\bar{\sigma}_2}, \quad N \to \infty. \tag{13.28}$$

Proof. By using the definitions of y_{12}, β_{13}, and β_{12} in Appendix A, we find that the approximation of the body and the near-end tails given in (13.25) takes the following form when $R_0 > 1$:

$$p_n^{(1)} \sim \frac{1}{\varphi(\beta_1)H_1(\beta_1)} \frac{1}{\bar{\sigma}_1} \varphi(y_1(n)), \quad R_0 > 1, \quad y_1(n) = o(\sqrt{N}), \quad N \to \infty. \tag{13.29}$$

Here, we have $\beta_1 \to +\infty$ as $N \to \infty$, since $R_0 > 1$. We conclude therefore from (13.1) that $H_1(\beta_1) \sim 1/\varphi(\beta_1)$. This shows that our approximation for $p_n^{(1)}$ given in (13.25) agrees with what has been derived in (7.2) in the parameter region where $R_0 > 1$.

Next, we use the definition of β_{13} in Appendix A to conclude that the approximation for the left tail given in (13.26) is as follows when $R_0 > 1$:

$$p_n^{(1)} \sim \frac{1}{H_1(\beta_1)} \frac{1}{\bar{\sigma}_2} \frac{\varphi(y_2(n))}{\varphi(\beta_2)}, \quad R_0 > 1, \quad n = o(N), \quad 1 \leq n, \quad N \to \infty. \tag{13.30}$$

As above, we have $\beta_1 \to +\infty$ and $H_1(\beta_1) \sim 1/\varphi(\beta_1)$ as $N \to \infty$, since $R_0 > 1$. We conclude therefore that the approximation claimed in (13.26) for the left body of $\mathbf{p}^{(1)}$ when $R_0 > 1$ agrees with what we have derived in (7.3).

Next, we note from (13.25) that the claim for the body of the distribution $\mathbf{p}^{(1)}$ in case $R_0 < 1$ is

$$p_n^{(1)} \sim \frac{1}{H_1(\beta_3)} \frac{1}{\bar{\sigma}_2} \frac{\varphi(y_2(n))}{\varphi(\beta_2)}, \quad R_0 < 1, \quad n = o(N), \quad 1 \leq n, \quad N \to \infty. \tag{13.31}$$

This expression agrees with the claim for the left tail given in (13.26), as it should. In this parameter region we have $\beta_3 \to -\infty$ as $N \to \infty$. By using the asymptotic behavior of H_1 in (13.1) we find that $H_1(\beta_3) \sim -1/\beta_3$. Therefore, we get

$$p_n^{(1)} \sim -\frac{\beta_3}{\bar{\sigma}_2} \frac{\varphi(y_2(n))}{\varphi(\beta_2)} = \frac{1 - R_0}{R_0} \frac{\varphi(y_2(n))}{\varphi(\beta_2)},$$

$$R_0 < 1, \quad n = o(N), \quad 1 \leq n, \quad N \to \infty, \tag{13.32}$$

using the definitions of β_3 and $\bar{\sigma}_2$ in Appendix A. This shows that the claimed approximations in both (13.25) and (13.26) in case $R_0 < 1$ agree asymptotically with the result given in (7.7).

We proceed to consider the transition region. After innocent changes, we find that the claimed uniform approximations in both (13.25) and (13.26) take the form

$$p_n^{(1)} \sim \frac{1}{H_1(\rho)} \frac{1}{\bar{\sigma}_3} \frac{\varphi(y_3(n))}{\varphi(\rho)}, \quad \rho = O(1), \quad n = o(N), \quad 1 \leq n, \quad N \to \infty. \quad (13.33)$$

This agrees with the approximation (7.26), since $H_1(\rho) = \Phi(\rho)/\varphi(\rho)$.

We note next that the approximation (13.27) in the left-most tail follows from the approximation given in (13.26) in the wider left tail by application of the relation $\varphi(y_2(2))/\varphi(\beta_2) \sim R_0^n$, which by (6.45) is valid for $n = o(\sqrt{N})$.

We note finally that the uniform approximation of the probability $p_1^{(1)}$ given in (13.28) is a special case of the uniform approximation of the left-most tail in (13.27), gotten by putting $n = 1$. $\square$

We note that the uniform approximation of $p_1^{(1)}$ given in (13.28) is consistent with the uniform approximation of $\sum_{n=1}^{N} \rho_n$ in (6.55), since $p_1^{(1)} = 1/\sum_{n=1}^{N} \rho_n$ and $\bar{\sigma}_2 = \sqrt{N}$.

By using the results in Chap. 7, we can conclude that the uniform approximation of the stationary distribution $\mathbf{p}^{(1)}$ makes a continuous transition from a geometric distribution when R_0 is distinctly below one to a distribution determined by a normal density function in its body and near-end tails when R_0 is increased to a value distinctly above one. The function H_1 serves to quantify this continuous transition.

13.3   Uniform Approximation of the Stationary Distribution $\mathbf{p}^{(0)}$

The stationary distribution $\mathbf{p}^{(0)}$ is approximated uniformly across the three parameter regions in this section. The first approximation in the theorem below gives a uniform approximation of the body and the near-end tails of this distribution, while the second approximation deals with the left tail, and the third one with the left-most tail. Finally, the fourth approximation gives a uniform approximation of the probability $p_1^{(0)}$. We refer to Appendix A for the notation that is used here.

Theorem 13.2 *The stationary distribution* $\mathbf{p}^{(0)}$ *of the SIS model has the following uniform approximation as* $N \to \infty$:

$$p_n^{(0)} \approx \frac{R_0}{r_0} \frac{1}{\frac{1}{2} \log N + H_0(\beta_{30})} \frac{1}{\varphi(\beta_{32})} \frac{\varphi(y_{12}(n))}{n},$$

$$y_1(n) = o(\sqrt{N}) \text{ when } R_0 > 1 \text{ as } N \to \infty,$$

$$\text{or } n \geq 1 \text{ and } n = o(N) \text{ when } \rho = O(1) \text{ as } N \to \infty,$$

$$\text{or } n \geq 1 \text{ and } n = o(N) \text{ when } R_0 < 1 \text{ as } N \to \infty, \quad (13.34)$$

$$p_n^{(0)} \approx \frac{\beta_3}{\beta_{13}} \frac{1}{\frac{1}{2} \log N + H_0(\beta_{10})} \frac{1}{\varphi(\beta_2)} \frac{\varphi(y_2(n))}{n}, \quad n = o(N), \quad 1 \leq n, \quad N \to \infty,$$

$$(13.35)$$

$$p_n^{(0)} \approx \frac{\beta_3}{\beta_{13}} \frac{1}{\frac{1}{2}\log N + H_0(\beta_{10})} \frac{R_0^n}{n}, \quad n = o(\sqrt{N}), \quad 1 \le n, \quad N \to \infty, \quad (13.36)$$

$$p_1^{(0)} \approx \frac{\rho}{\beta_{13}} \frac{1}{\frac{1}{2}\log N + H_0(\beta_{10})}, \quad N \to \infty. \quad (13.37)$$

Proof. The uniform approximation of the body and the near-end tails of the distribution $\mathbf{p}^{(0)}$ claimed in (13.34) takes the following form in the parameter region where $R_0 > 1$:

$$p_n^{(0)} \approx \frac{\sqrt{R_0}}{\frac{1}{2}\log N + H_0(\beta_3)} \frac{1}{\varphi(\beta_3)} \frac{\varphi(y_1(n))}{n},$$

$$R_0 > 1, \quad y_1(n) = o(\sqrt{N}), \quad 1 \le n, \quad N \to \infty. \quad (13.38)$$

Here, $\beta_3 \to +\infty$ as $N \to \infty$. We conclude therefore from (13.2) that $H_0(\beta_3) \sim 1/(\beta_3\varphi(\beta_3))$. Furthermore, we note that the term $\frac{1}{2}\log N$ is of smaller magnitude than $H_0(\beta_3)$, and that $n = \bar{\mu}_1 + \bar{\sigma}_1 y_1(n) \sim \bar{\mu}_1$ for $y_1(n) = o(\sqrt{N})$. Thus we get

$$p_n^{(0)} \approx \frac{\sqrt{R_0}\beta_3\varphi(\beta_3)}{\varphi(\beta_3)\bar{\mu}_1}\varphi(y_1(n)) = \frac{1}{\bar{\sigma}_1}\varphi(y_1(n)),$$

$$R_0 > 1, \quad y_1(n) = o(\sqrt{N}), \quad N \to \infty. \quad (13.39)$$

This agrees with the approximation given in (9.2). Thus, the uniform approximation claimed in (13.34) for the body of $\mathbf{p}^{(0)}$ when $R_0 > 1$ agrees with what has been derived in Chap. 9.

Next, we consider the claim that (13.35) gives a uniform approximation of the left tail of the stationary distribution $\mathbf{p}^{(0)}$. In the parameter region $R_0 > 1$ it takes the following form:

$$p_n^{(0)} \approx \frac{\beta_3}{\beta_1} \frac{1}{\frac{1}{2}\log N + H_0(\beta_1)} \frac{1}{\varphi(\beta_2)} \frac{\varphi(y_2(n))}{n},$$

$$R_0 > 1, \quad n = o(N), \quad 1 \le n, \quad N \to \infty. \quad (13.40)$$

Here, $\beta_1 \to +\infty$ as $N \to \infty$. Using (13.2), we conclude that $H_0(\beta_1) \sim 1/(\beta_1\varphi(\beta_1))$, and that the term $\frac{1}{2}\log N$ can be ignored compared to the term $H_0(\beta_1)$. We get

$$p_n^{(0)} \approx \beta_3 \frac{\varphi(\beta_1)}{\varphi(\beta_2)} \frac{\varphi(y_2(n))}{n}, \quad R_0 > 1, \quad n = o(N), \quad 1 \le n, \quad N \to \infty. \quad (13.41)$$

This agrees with the approximation (9.3) derived in Chap. 9.

We now consider the two approximations claimed in (13.34) and (13.35) in the parameter region where $R_0 < 1$. The approximation claimed in (13.34) can then be written

$$p_n^{(0)} \approx \frac{1}{\frac{1}{2}\log N + H_0(\rho)} \frac{1}{n} \frac{\varphi(y_2(n))}{\varphi(\beta_2)}, \quad R_0 < 1, \quad n = o(N), \quad 1 \le n, \quad N \to \infty.$$

(13.42)

The approximation in (13.35) takes the same form, as it should. In this parameter region we have $\rho \to -\infty$ as $N \to \infty$. By using (13.2) we get $H_0(\rho) \sim -\log|\rho|$ and

$$\frac{1}{2}\log N + H_0(\rho) \sim \log \frac{\sqrt{N}}{|\rho|} = \log \frac{1}{1 - R_0}, \quad R_0 < 1, \quad N \to \infty. \qquad (13.43)$$

We get therefore

$$p_n^{(0)} \approx \frac{1}{\log(1/(1 - R_0))} \frac{1}{n} \frac{\varphi(y_2(n))}{\varphi(\beta_2)}, \quad R_0 < 1, \quad n = o(N), \quad 1 \le n, \quad N \to \infty.$$

(13.44)

This agrees with the approximation given in (9.10).

We proceed to consider the approximations claimed in (13.34) and (13.35) in the transition region. After innocent changes, both of them can be written

$$p_n^{(0)} \approx \frac{1}{\frac{1}{2}\log N + H_0(\rho)} \frac{1}{n} \frac{\varphi(y_3(n))}{\varphi(\rho)}, \quad \rho = O(1), \quad n = o(N), \quad 1 \le n, \quad N \to \infty.$$

(13.45)

This agrees with the result given in (9.13). Thus we have proved that the uniform approximations of the stationary distribution $\mathbf{p}^{(0)}$ given in (13.34) and (13.35) are valid.

In similarity to the case with the stationary distribution $\mathbf{p}^{(1)}$ dealt with in the preceding section, we note that the uniform approximation of the left-most tail given in (13.36) follows from the uniform approximation given for the wider left tail in (13.35) by using the approximation $\varphi(y_2(n))/\varphi(\beta_2) \sim R_0^n$ given in (6.45) when $n = o(\sqrt{N})$.

We note finally that the uniform approximation of the probability $p_1^{(0)}$ given in (13.37) follows from the uniform approximation for the left-most tail in (13.36) by setting $n = 1$, and recognizing that $R_0\beta_3 = \rho$. □

By using the results in Chap. 9, we can conclude that the uniform approximation of the stationary distribution $\mathbf{p}^{(0)}$ makes a continuous transition from a log series distribution when R_0 is distinctly below one to a distribution determined by a normal density function in its body and near-end tails when R_0 is increased to a value distinctly above one. The function H_0 serves to quantify this continuous transition.

13.4   Uniform Approximation of the Quasi-stationary Distribution q

In this section we reach a final point in our search for knowledge about the quasi-stationary distribution of the SIS model. The information about the behavior of the quasi-stationary distribution in each of the three parameter regions that was established in Chap. 11 will here be used to derive uniform approximations that are valid across all three of the parameter regions. These uniform approximations provide improvements over the three specific approximations given in Chap. 11.

In a close parallel to the presentations of the two previous sections, we give our results in a theorem that contains four uniform approximations that are valid over different n-values. The first of these approximations gives a uniform approximation of the quasi-stationary distribution in its body and near-end tails. The second approximation, in contrast, approximates the quasi-stationary distribution in the left tail, and the third one in the left-most tail. The final result is a uniform approximation of the important probability q_1, which carries information about the expected time to extinction from quasi-stationarity for the SIS model. We use notation listed in Appendix A.

Theorem 13.3 *The quasi-stationary distribution* **q** *of the SIS model has the following uniform approximation as* $N \to \infty$:

$$q_n \approx \frac{1}{r_0} \frac{1}{1+\rho H(\rho)} \frac{1}{n} \left(1 - \frac{1}{[R(N,\rho)]^n}\right) \frac{\varphi(y_{12}(n))}{\varphi(\beta_{02})},$$

$$y_1(n) = o(\sqrt{N}) \text{ when } R_0 > 1 \text{ as } N \to \infty,$$

$$\text{or } n \geq 1 \text{ and } n = o(N) \text{ when } \rho = O(1) \text{ as } N \to \infty,$$

$$\text{or } n \geq 1 \text{ and } n = o(N) \text{ when } R_0 < 1 \text{ as } N \to \infty, \quad (13.46)$$

$$q_n \approx \frac{1}{R_0} \frac{\rho}{\beta_{10}} \frac{1}{1+\beta_{10}H(\beta_{10})} \frac{1}{n} \left(1 - \frac{1}{[R(N,\rho)]^n}\right) \frac{\varphi(y_2(n))}{\varphi(\beta_2)},$$

$$n = o(N), \quad 1 \leq n, \quad N \to \infty, \quad (13.47)$$

$$q_n \approx \frac{\rho}{\beta_{10}} \frac{1}{1+\beta_{10}H(\beta_{10})} \left(1 - \frac{1}{[R(N,\rho)]^n}\right) \frac{R_0^{n-1}}{n},$$

$$n = o(\sqrt{N}), \quad 1 \leq n, \quad N \to \infty, \quad (13.48)$$

$$q_1 \approx \frac{\rho}{\beta_{10}} \frac{1}{1+\beta_{10}H(\beta_{10})} \frac{R(N,\rho)-1}{R(N,\rho)}, \quad N \to \infty. \quad (13.49)$$

Proof. We show first that the approximations given in (13.46) and (13.47) are valid in the parameter region where $R_0 > 1$. Here we have $r_0 = \sqrt{R_0}$, $y_{12}(n) = y_1(n)$, and $\beta_{02} = \beta_0 = \rho$. The approximation of the body in (13.46) can therefore be written as follows when $R_0 > 1$:

$$q_n \approx \frac{1}{\sqrt{R_0}} \frac{1}{1 + \rho H(\rho)} \frac{1}{n} \left(1 - \frac{1}{[R(N,\rho)]^n}\right) \frac{\varphi(y_1(n))}{\varphi(\rho)},$$

$$R_0 > 1, \quad y_1(n) = o(\sqrt{N}), \quad N \to \infty. \quad (13.50)$$

Here, $\rho \to +\infty$ as $N \to \infty$. Hence, applying (13.3), we find $H(\rho) \sim 1/(\rho^2 \varphi(\rho))$ and $1 + \rho H(\rho) \sim 1/(\rho \varphi(\rho))$. Furthermore, we have $n \sim \bar{\mu}_1$ and, using (13.4), $R(N,\rho) \sim R_0$. Clearly, $n \to \infty$ as $N \to \infty$, so $1/R_0^n$ is negligible. Thus,

$$q_n \approx \frac{1}{\sqrt{R_0}} \frac{\rho \varphi(\rho)}{\bar{\mu}_1 \varphi(\rho)} \varphi(y_1(n)) = \frac{1}{\bar{\sigma}_1} \varphi(y_1(n)), \quad R_0 > 1, \quad y_1(n) = o(\sqrt{N}), \quad N \to \infty.$$

$$(13.51)$$

This shows that the approximation of the body and the near-end tails of the quasi-stationary distribution given by (13.46) agrees asymptotically with the approximation given in (11.1) when $R_0 > 1$.

Next we consider the approximation of the left tail of the quasi-stationary distribution given in (13.47). For $R_0 > 1$ we have $\beta_{10} = \beta_1$. Therefore, the approximation claimed in (13.47) can be expressed as follows:

$$q_n \approx \frac{\rho}{\beta_1 R_0} \frac{1}{1 + \beta_1 H(\beta_1)} \frac{1}{n} \left(1 - \frac{1}{[R(N,\rho)]^n}\right) \frac{\varphi(y_2(n))}{\varphi(\beta_2)},$$

$$R_0 > 1, \quad n = o(N), \quad 1 \le n, \quad N \to \infty. \quad (13.52)$$

Here, we have $\beta_1 \to +\infty$ as $N \to \infty$. Thus, using the asymptotic approximation of $H(x)$ in (13.3) as $x \to +\infty$, we get $H(\beta_1) \sim 1/(\beta_1^2 \varphi(\beta_1))$, and $1 + \beta_1 H(\beta_1) \sim 1/(\beta_1 \varphi(\beta_1))$. Furthermore, the asymptotic approximation of $R(N,\rho)$ as $\rho \to \infty$ given in (13.4) gives $R(N,\rho) \sim R_0$. Thus, writing $\rho = (R_0 - 1)\sqrt{N}$, we get

$$q_n \approx \frac{(R_0 - 1)\sqrt{N}}{\beta_1 R_0} \beta_1 \varphi(\beta_1) \frac{1}{n} \left(1 - \frac{1}{R_0^n}\right) \frac{\varphi(y_2(n))}{\varphi(\beta_2)},$$

$$R_0 > 1, \quad n = o(N), \quad 1 \le n, \quad N \to \infty. \quad (13.53)$$

For small n-values, satisfying $n = o(\sqrt{N})$, we use the approximation in (6.45), namely $\varphi(y_2(n))/\varphi(\beta_2) \sim R_0^n$, to conclude that the claimed approximation of the left-most tail coincides with the approximation (11.4) derived in Chap. 11. For larger n-values, with $n > N^{1/4}$, we note that $R_0^n \to \infty$ as $N \to \infty$. The term $1/R_0^n$ can therefore be ignored compared to one. Therefore, the above expression for q_n is asymptotically equal to the one derived in Chap. 11, and given in (11.3). We conclude that the approximation of the left tail of the quasi-stationary distribution claimed in (13.47) agrees asymptotically with the result that has been derived in Chap. 11 in case $R_0 > 1$.

Next we consider the parameter region where $R_0 < 1$. The approximation claimed in (13.46) can then be written as follows, since $r_0 = R_0$, $y_{12}(n) = y_2(n)$, and $\beta_{02} = \beta_2$:

$$q_n \approx \frac{1}{R_0} \frac{1}{1+\rho H(\rho)} \frac{1}{n} \left(1 - \frac{1}{[R(N,\rho)]^n}\right) \frac{\varphi(y_2(n))}{\varphi(\beta_2)},$$

$$R_0 < 1, \quad n = o(N), \quad 1 \le n, \quad N \to \infty. \quad (13.54)$$

The approximation of the left tail claimed in (13.47) obeys the same expression, as it should, since $\beta_{10} = \beta_0 = \rho$. In this parameter region we have $\rho \to -\infty$ as $N \to \infty$. We use the asymptotic approximation of $H(\rho)$ given in (13.3), which says that $H(\rho) \sim -1/\rho + 2/\rho^3$. This gives $1 + \rho H(\rho) \sim 2/\rho^2$. Furthermore, using the asymptotic approximation of $R(N,\rho)$ given in (13.4), we get $R(N,\rho) \sim 1 - 2/(\rho\sqrt{N}) = 1 + 2/((1-R_0)N)$. Therefore,

$$\log[R(N,\rho)]^n = n \log R(N,\rho) \sim n \log\left(1 + \frac{2}{(1-R_0)N}\right) \sim \frac{2n}{(1-R_0)N},$$

$$R_0 < 1, \quad n = o(N), \quad 1 \le n, \quad N \to \infty, \quad (13.55)$$

and hence

$$[R(N,\rho)]^n = \exp(n \log R(N,\rho)) \sim 1 + \frac{2n}{(1-R_0)N},$$

$$R_0 < 1, \quad n = o(N), \quad 1 \le n, \quad N \to \infty. \quad (13.56)$$

This implies that

$$1 - \frac{1}{[R(N,\rho)]^n} = \frac{[R(N,\rho)]^n - 1}{[R(N,\rho)]^n} \sim \frac{2n}{(1-R_0)N},$$

$$R_0 < 1, \quad n = o(N), \quad 1 \le n, \quad N \to \infty. \quad (13.57)$$

This means that the claim in (13.54) can be written

$$q_n \approx \frac{1}{R_0} \frac{\rho^2}{2} \frac{1}{n} \frac{2n}{(1-R_0)N} \frac{\varphi(y_2(n))}{\varphi(\beta_2)} = \frac{1-R_0}{R_0} \frac{\varphi(y_2(n))}{\varphi(\beta_2)},$$

$$R_0 < 1, \quad n = o(N), \quad 1 \le n, \quad N \to \infty, \quad (13.58)$$

where we have inserted $\rho^2 = (1 - R_0)^2 N$. This expression agrees with the result given in (11.12). It follows that the approximations claimed in (13.46) and (13.47) agree asymptotically with the approximation of the quasi-stationary distribution for $R_0 < 1$ given in Chap. 11.

We proceed to consider the transition region. We then have $R_0 \sim 1$, $r_0 \sim 1$, $y_{12}(n) \sim y_3(n)$, and $\beta_{02} \sim \rho$. By applying corresponding innocent changes, we find that the approximation claimed in (13.46) is asymptotically equal to

$$q_n \approx \frac{1}{\varphi(\rho)} \frac{1}{1 + \rho H(\rho)} \frac{1}{n} \left(1 - \frac{1}{[R(N, \rho)]^n} \right) \varphi(y_3(n)),$$

$$\rho = O(1), \quad n = o(N), \quad 1 \leq n, \quad N \to \infty. \quad (13.59)$$

It is readily seen that the approximation claimed in (13.47) is asymptotically equal to the same expression in the transition region, since $\beta_{10} \sim \rho$, $y_2(n) \sim y_3(n)$, and $\beta_2 \sim \rho$. This claimed approximation of the quasi-stationary distribution in the transition region agrees with the result in (11.26).

We note furthermore that the approximation of the left-most tail of the quasi-stationary distribution $\mathbf{q}$ given in (13.48) follows from the approximation of the wider left tail in (13.47) by application of the approximation of $\varphi(y_2(n))/\varphi(\beta_2) \sim R_0^n$ given in (6.45), and valid for $n = o(\sqrt{N})$.

Finally, the uniform approximation of the single probability q_1 given in (13.49) follows directly from the uniform approximation of the left-most tail in (13.48) by setting $n = 1$.

We conclude that the uniform approximations of the quasi-stationary distribution that we have given in (13.46)–(13.49) agree asymptotically in each of the three parameter regions with the results that have been derived in Chap. 11 in each parameter region separately.   □

By using the results in Chap. 11, we can conclude that the uniform approximation of the quasi-stationary distribution $\mathbf{q}$ makes a continuous transition from a geometric distribution when R_0 is distinctly below one to a distribution determined by a normal density function in its body and near-end tails when R_0 is increased to a value distinctly above one. The function H serves to quantify this continuous transition. It is interesting to note that the end-points of the transition for the quasi-stationary distribution $\mathbf{q}$ are similar to the end-points for the transition of the stationary distribution $\mathbf{p}^{(1)}$. However, the paths travelled by the two distributions between the two end-points differ. This is equivalent to saying that the two distributions differ in the transition region.

13.5   Uniform Approximation of the Time to Extinction

The time to extinction is a random variable, whose distribution depends on the initial distribution of infected individuals. We deal with two cases. In one of them, the initial distribution equals the quasi-stationary distribution, and in the other one we start with exactly one infected individual. The time to extinction in the first case, τ_Q, is a measure of the persistence of an established infection. It has an exponential distribution, whose expectation can be used to establish a persistence threshold, as shown in the next chapter. In the second case, the time to extinction is denoted τ_1. Its distribution is more complicated; we restrict our attention to its expectation.

Uniform approximations of the two extinction time expectations are given in the following theorem. As before, we use the notation in Appendix A.

Theorem 13.4 *The expected time to extinction for the SIS model has the following uniform approximations in the two cases when initially quasi-stationarity holds* $(E\tau_Q)$, *and when initially there is one infected individual* $(E\tau_1)$:

$$E\tau_Q \approx \frac{1}{\mu} \frac{R(N,\rho)}{R(N,\rho)-1} \frac{\beta_{10}}{\rho} \left(1+\beta_{10}H(\beta_{10})\right), \quad N \to \infty, \tag{13.60}$$

$$E\tau_1 \approx \frac{1}{\mu} \frac{\beta_{13}}{\rho} \left(\frac{1}{2}\log N + H_0(\beta_{10})\right), \quad N \to \infty. \tag{13.61}$$

Proof. We recall from (3.42) that the expected time to extinction from quasi-stationarity, $E\tau_Q$, is determined from the probability q_1 by the relation

$$E\tau_Q = \frac{1}{\mu_1 q_1}. \tag{13.62}$$

The uniform approximation of $E\tau_Q$ in (13.60) follows from this relation by using the uniform approximation of q_1 in (13.49) and the fact that $\mu_1 = \mu$ for the SIS model.

Similarly we find from (3.44) that the expected time to extinction from the state one, $E\tau_1$, is determined from the probability $p_1^{(0)}$ by the relation

$$E\tau_1 = \frac{1}{\mu_1 p_1^{(0)}}. \tag{13.63}$$

The uniform approximation of $E\tau_1$ in (13.61) follows from this relation by using the uniform approximation of $p_1^{(0)}$ in (13.37), and the fact that $\mu_1 = \mu$ for the SIS model. □

Chapter 14
Thresholds for the SIS Model

As mentioned in the introduction, threshold concepts are the most powerful results that mathematical modelling can contribute to theoretical epidemiology. They lead to threshold functions that partition the parameter space into subsets in which the model solutions behave in qualitatively different ways. Early threshold results are all based on deterministic models. For the deterministic version of the SIS model, one studies the proportion of infected individuals as time approaches infinity. To describe the ultimate value of this proportion, one defines a parameter conventionally denoted R_0 and called the basic reproduction ratio. If R_0 is larger than one, then any positive proportion of infected individuals leads to a positive endemic level of infection, while any existing infection will ultimately disappear if $R_0 \leq 1$. Thus, the threshold value $R_0 = 1$ defines a boundary in parameter space between those communities that can support a positive endemic infection level and those that do not.

Our emphasis on fully stochastic models leads us naturally to seek counterparts to the deterministic threshold in the stochastic setting. However, thresholds for the corresponding stochastic model can not be based on the same distinction between survival or extinction of the infection, since the stochastic model predicts ultimate extinction of any initial infection. We use instead the time to extinction as a measure that separates those communities in which any established infection lasts for a long time from those where extinction occurs quickly. Indeed, this measure allows us to define two thresholds, namely an invasion threshold and a persistence threshold. These thresholds allow us to respond to the following two questions:

(1) Will the introduction of a small amount of infection into an infection-free community spread so that the infection lasts for a long time?
(2) Will an established endemic infection level persist for a long time?

To proceed we quantify first what is meant by a "long" time. We agree to relate any waiting time to the natural time constant $1/\mu$ for the model, where μ is the recovery rate per individual. Next, we regard a waiting time to be long if it is one or two orders of magnitude larger than the natural time constant for the model.

I. Nåsell, *Extinction and Quasi-stationarity in the Stochastic Logistic SIS Model*,
Lecture Notes in Mathematics 2022, DOI 10.1007/978-3-642-20530-9_14,
© Springer-Verlag Berlin Heidelberg 2011

Thirdly, we agree to measure any extinction time by its expectation. Finally, it is obvious that the time to extinction, which is a stochastic variable, certainly depends on the initial distribution of infected individuals. Thus, we start from one infected individual when we deal with the invasion threshold, and from the quasi-stationary distribution when we are concerned with the persistence threshold. The invasion threshold is denoted by R_{0I} and the persistence threshold by R_{0P}. Pulling these facts together, we are led to the following definitions:

Definition 1. The stochastic SIS model has its *invasion threshold* R_{0I} at that value of the basic reproduction ratio R_0 for which the expected time to extinction $E\tau_1$ from the state one equals K/μ, where K equals 10 or 100. In other words, R_{0I} is found by solving $E\tau_1 = K/\mu$ for R_0 as a function of N and K.

Definition 2. The stochastic SIS model has its *persistence threshold* R_{0P} at that value of the basic reproduction ratio R_0 for which the expected time to extinction $E\tau_Q$ from the quasi-stationary distribution equals K/μ, where K equals 10 or 100. In other words, R_{0P} is found by solving $E\tau_Q = K/\mu$ for R_0 as a function of N and K.

It follows from these definitions that both of the thresholds depend on N and K. The dependence on N is typical of the stochastic model. It does not appear in the deterministic model threshold. This reflects the fact that the spaces of essential parameters needed for describing stochastic and deterministic models differ. For the stochastic model we consider the two-dimensional space where R_0 and N take their values, while the one-dimensional space of values of R_0 suffices for the deterministic model. This is consistent with the fact that the deterministic model can be derived from the stochastic one by a scaling by N, followed by taking the limit as $N \to \infty$. Thus, N disappears from the scene as we go from the stochastic to the deterministic model.

We note that our threshold definitions are closely related to the so-called critical community size introduced by Bartlett [12]. Any one of the two thresholds is for given value of K determined by a function of the population size N. The inverse of this function gives the population size as a function of R_0. This function is denoted the critical population size by Bartlett.

The definitions of invasion and persistence thresholds given by Nåsell [43, 44] allow the constant K to be replaced by a nondecreasing function $f(N)$. This slightly more general definition will not be used here. Numerical evaluations of the two thresholds are given in Nåsell [44] for $K = 10$. These evaluations lead to the conjectures that the invasion threshold for $K = 10$ can for large N be approximated by $R_{0I} \approx 1 + 1.6/\sqrt{N}$, and the persistence threshold by $R_{0P} \approx 0.9 + 18/N$. It was furthermore left as open problems in the 1996 paper to prove the validity of these conjectures. We proceed to state the main result of the present chapter, namely approximations of the two threshold functions. We shall then respond to the open problem formulated in the 1996 paper by showing that the results given here lead to slight numerical modifications of the conjectures quoted above.

We derive below explicit approximations of the two threshold functions that show their dependence on N and K. The approximations require both N and K to be large. In addition, we assume that N is much smaller than $\exp(2K)$. This last requirement is well satisfied for the K-values 10 and 100 if $N << 4.9 \cdot 10^8$, since $\exp(20) \approx 4.9 \cdot 10^8$ and $\exp(200) \approx 7.2 \cdot 10^{86}$. The results are as follows:

$$R_{0I} \approx 1 + \frac{1}{\sqrt{N}} H_0^{-1}(K - \log \sqrt{N}), \quad N << \exp(2K), \quad N \to \infty, \quad K \to \infty. \quad (14.1)$$

and

$$R_{0P} \approx 1 - \frac{1}{K} + \frac{2K}{N}, \quad N << \exp(2K), \quad N \to \infty, \quad K \to \infty. \quad (14.2)$$

Here, H_0^{-1} denotes the inverse of the function H_0.

To derive the above approximation of the invasion threshold, we need to solve the equation $E\tau_1 = K/\mu$ for R_0 as a function of N and K. A preliminary investigation shows that the solution is found in the transition region where $\rho = O(1)$. The specific equation to be solved can therefore be written as follows:

$$\log \sqrt{N} + H_0(\rho) \sim K, \quad \rho = O(1), \quad N \to \infty, \quad (14.3)$$

using the expression for $E\tau_1$ given in (12.16). This equation has the solution $\rho = H_0^{-1}(K - \log \sqrt{N})$. By using the relation $\rho = (R_0 - 1)\sqrt{N}$ we arrive at the approximation of the invasion threshold given in (14.1).

To derive the approximation of the persistence threshold given in (14.2), we solve the equation $E\tau_q = K/\mu$ for R_0 as a function of N and K. In this case, a preliminary investigation shows that the solution is found in the parameter region where $R_0 < 1$. We determine the specific equation to be solved by starting out with the uniform approximation for the expected time to extinction from the quasi-stationary distribution given in (13.60). Thus, we get

$$\frac{R(N,\rho)}{R(N,\rho) - 1} \frac{\beta_{10}}{\rho} (1 + \beta_{10} H(\beta_{10})) \approx K, \quad R_0 < 1, \quad (14.4)$$

where $R = R(N,\rho)$. In the parameter region $R_0 < 1$ we have $\beta_{10} = \beta_0 = \rho$, using the definition of β_{10} in Appendix A. Thus, the equation to be solved for R_0 can be written

$$\frac{R(N,\rho)}{R(N,\rho) - 1} (1 + \rho H(\rho)) \approx K, \quad R_0 < 1. \quad (14.5)$$

Table 14.1 The function
$\rho_I(N) = H_0^{-1}(K - \log \sqrt{N})$ is
shown to vary slowly with N,
for $K = 10$ and $10 \leq N \leq 10^4$

N	$\rho_I(N)$
10^1	1.86
10^2	1.77
10^3	1.66
10^4	1.52

By using the definition of the function R in (11.19) this becomes

$$R(N,\rho)H(\rho) \approx \frac{K}{\sqrt{N}}, \quad R_0 < 1. \tag{14.6}$$

Since by (13.4) we have $R(N,\rho) \sim 1$ in the parameter region where $R_0 < 1$ we get
the relation

$$H(\rho) \approx \frac{K}{\sqrt{N}}, \quad R_0 < 1, \quad N \to \infty. \tag{14.7}$$

By using the asymptotic approximation of $H(y)$ as $y \to -\infty$ in (13.3) we are led to
the equation $-1/\rho + 2/\rho^3 \sim K/\sqrt{N}$, whose solution can be written $\rho \sim -\sqrt{N}/K +
2K/\sqrt{N}$. By introducing $R_{0P} = 1 + \rho/\sqrt{N}$, we are led to the approximation for the
persistence threshold given in (14.2).

It remains to consider the conjectures advanced in the 1996 paper concerning
the two threshold functions. Concerning the invasion threshold we note that the
conjecture based on numerical evaluations was that $R_{0I} \approx 1 + 1.6/\sqrt{N}$ for $K = 10$,
while the approximation in (14.1) can be written $R_{0I} \approx 1 + \rho_I(N)/\sqrt{N}$, where
$\rho_I(N) = H_0^{-1}(K - \log \sqrt{N})$. These expressions are clearly different. However, the
restriction $N << \exp(2K)$ that holds for our analytic approximation in (14.1)
implies that $\rho_I(N)$ is a function of N that varies slowly with N. We illustrate this by
giving numerical values of this function for a few N-values and $K = 10$ in Table 14.1.
We have no explicit expression for the function H_0^{-1}, but numerical evaluations are
straightforward from the definition in (8.10) of the function H_0. (We use $\rho_b = -3$.)
The table indicates that the function $\rho_I(N)$ may be approximated by a constant
of the order of 1.6 or 1.7 when N takes values from 10 to 10^4 and $K = 10$. The
slow variation of $\rho_I(N)$ for $K = 10$ shows the extent to which the conjecture based
on numerical evaluations contained in the 1996 paper differs from the analytical
approximation of the invasion threshold contained in (14.1).

The approximation of the persistence threshold based on the numerical evaluations reported in the 1996 paper was $R_{0P} \approx 0.9 + 18/N$ for large N-values and
$K = 10$. This deviates slightly from the analytical approximation of the persistence
threshold given (14.2), since the latter one gives $R_{0P} \approx 0.9 + 20/N$ for $K = 10$.

We conclude from these comparisons that the analytical approximations of the
threshold functions that are presented in (14.1) and (14.2) support the course
approximations given in the 1996 paper in a qualitative way. Thus, they respond to
the demand for confirmation by analytic arguments pronounced in the 1996 paper.

We summarize our discussion of threshold functions for the stochastic SIS model in the following way:

We note first of all that two threshold functions can be defined for the stochastic model, namely an invasion threshold and a persistence threshold. This is distinct from the situation for the deterministic SIS model, where only one threshold appears. Among the two thresholds, the invasion threshold is higher than the persistence threshold. This means that a larger effort is required to eradicate an established infection than to prevent an infection to establish itself. The difference between the two threshold functions is seen in two ways. For one, we note that the invasion threshold approaches the value one as N approaches infinity, while the persistence threshold approaches $1 - 1/K$, which is smaller than one. This is also reflected in the result that the equation $E\tau_1 = K/\mu$ has its solution in the transition region where $\rho = O(1)$ as $N = \rightarrow \infty$, while the equation $E\tau_Q = K/\mu$ has its solution in the parameter region where R_0 is distinctly below one. In addition, we note that the second term in the approximation of the invasion threshold is a decreasing function of N which is proportional to $1/\sqrt{N}$, while the second term in the approximation of the persistence threshold is a decreasing function of N which is proportional to $1/N$. Thus, the difference between the two thresholds is an increasing function of N.

We end this chapter by pointing out a weakness with the definition of persistence thresholds. The definition deals with persistence of an established infection. The persistence depends of course on the distribution of infected individuals at the time when we wish to measure the persistence. It is therefore important to acquire information about this distribution. We consider a situation where it is known that an infection exists, and that the community has been exposed to constant environmental influences for a long time. One can then use a mathematical result that says that the distribution of infected individuals is approximated by the quasi-stationary distribution, regardless of the distribution that existed a long time ago. This result is used in the definition of the persistence threshold. Its validity is however not universal. The result holds well in the parameter region where $R_0 > 1$ and quite well also in the transition region, but less so in the parameter region where $R_0 < 1$. This indicates a weak point in our definition, since our above arguments show that the persistence threshold for the SIS model is found in the parameter region where $R_0 < 1$. We leave it as an open problem to discuss this weakness.

Chapter 15
Concluding Comments

The story about the the extinction time and the quasi-stationary distribution of the SIS model has come to an end. The final results are given in terms of uniform approximations of the quasi-stationary distribution in Theorem 13.3, and uniform approximations of the expected times to extinction from the state one and from quasi-stationarity in Theorem 13.4. We remind the reader that these final approximations are uniform across the three parameter regions in which qualitatively different behaviors are observed.

We use this chapter to compare the approximations that we have derived with numerical evaluations of the quantities that are approximated. We shall also comment on how the results that we present compare with approximations that were available previously.

The quasi-stationary distribution is plotted together with one or two of its uniform approximations in Figs. 15.1 and 15.2. Note that the same parameter values are used as in Fig. 4.1. In Fig. 15.1 we give three plots for each of two different values of R_0. The vertical scale is linear in the first plot for each value of R_0, where the body of the distribution is shown together with its approximation (13.46). The vertical scale is logarithmic in the other two plots for each R_0. In the second plot for each R_0 the whole distribution is plotted, together with the approximation of the body, and part of the approximation (13.47) of the left tail. In the third plot for each R_0 we concentrate on the behavior of the left tail. Here, we plot the distribution q together with its two approximations. The plots that cover the body of the distribution indicate that the approximation (13.46) agrees reasonably well with the body of the quasi-stationary distribution. The plots that cover the whole distribution, with n-values from 1 to N, show that the right tail is strongly overestimated by the approximation of the body, and even more so by the approximation of the left tail. This is no surprise. The conditions on n given in (13.46) and (13.47) show that both of these expressions are approximations of q_n only for limited ranges of n-values that definitely exclude the right tail. The third plot for each of the two R_0-values in Fig. 15.1 illustrate an important result. For $R_0 = 1.5$ we see that the approximations of the left tail and of the body are clearly different, and that the first

I. Nåsell, *Extinction and Quasi-stationarity in the Stochastic Logistic SIS Model*, Lecture Notes in Mathematics 2022, DOI 10.1007/978-3-642-20530-9_15, © Springer-Verlag Berlin Heidelberg 2011

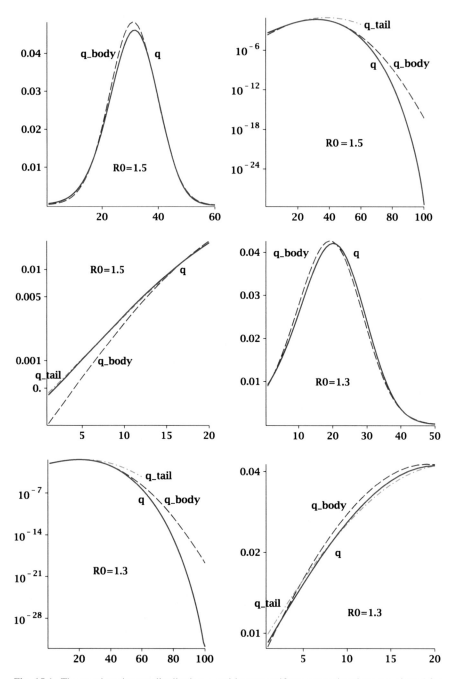

Fig. 15.1 The quasi-stationary distribution q and its two uniform approximations are plotted for $N = 100$ and two different R_0-values. Note that all the distributions are discrete. Thus, the *curve* for e.g. q gives the probability q_k as a function of k

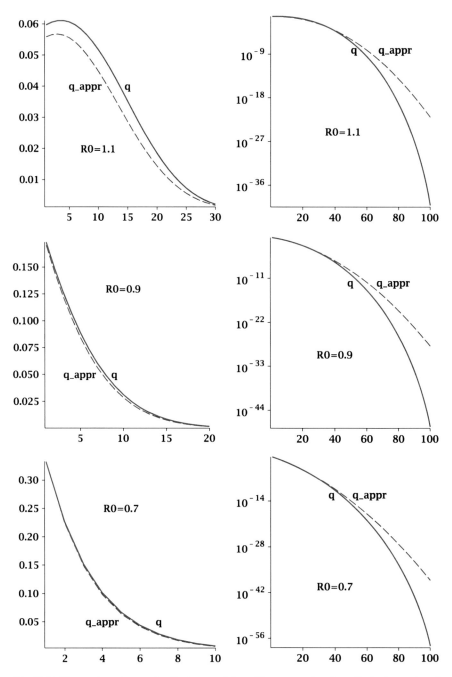

Fig. 15.2 The quasi-stationary distribution q and the uniform approximation of its body are plotted for $N = 100$ and three different R_0-values. Note that all the distributions are discrete. Thus, the *curve* for e.g. q gives the probability q_k as a function of k

one agrees rather well with the quasi-stationary distribution q, but that the second one gives an underestimate. However, for the lower value of R_0, 1.3, we see that the approximation of the left tail and the approximation of the body are rather close, and also quite close to the numerically evaluated left tail of the quasi-stationary distribution q. This is in line with the conditions on n spelled out in (13.46) and (13.47). Indeed, the body of the distribution agrees asymptotically with its left tail in the transition region. By adopting the rule of thumb that the value $\rho = 3$ forms the boundary between the transition region and the region where R_0 is distinctly larger than one, we would expect this behavior, since $R_0 = 1.3$ corresponds to the ρ-value 3 when $N = 100$, while the value $R_0 = 1.5$ gives $\rho = 5$ and so winds up outside the transition region.

The plots given in Fig. 15.2 all deal with values of R_0 that are equal to or smaller than 1.1. By previous arguments, this means that we are in the transition region or perhaps in the region where R_0 is distinctly below one. The two approximations of the quasi-stationary distribution q given in (13.46) and (13.47) then agree asymptotically. We therefore plot only one approximation for these cases. Two plots are given for each of three different values of R_0. The vertical scale is linear in one of the plots, and logarithmic in the other one. The plot with linear vertical scale shows the body and the left tail of the distribution, while the other one indicates the behavior in the right tail. We conclude that the body and the left tail are reasonably well approximated, but that the right tail is seriously overestimated by our approximation. The numerical evaluations behind the plots in Figs. 15.1 and 15.2 have been done with the Maple procedures included in Appendix B.

We note that we have limited ourselves to deriving approximations of the body and of the left tail of the quasi-stationary distribution, and that no effort has been devoted to the right tail. The reason for including the left tail but not the right tail in our study is that the left tail carries information about the time to extinction from quasi-stationarity, and that a similar importance of the right tail is missing. However, if an interest in the right tail would arise, then we note that the methods that we have used can be applied also to this case. The starting point for such an investigation would be the approximation of ρ_n in (6.10), with $\alpha < 1$.

The approximation of the expected time to extinction from quasi-stationarity, $E\tau_Q$, is compared with numerically evaluated values in Fig. 15.3. The figure indicates that the uniform approximation is superior to the three approximations that are available in each of the three parameter regions. The similarity with Fig. 6.2 is noted.

The approximations of the quasi-stationary distribution of the SIS model can be compared with the results given by Nåsell [47], who dealt with the quasi-stationary distribution of the more general Verhulst model. By adjusting its parameters to the SIS model that we deal with here, we find that the approximations given here are extensions of the results from 2001. Thus, the approximation of the body of the quasi-stationary distribution given in (11.1) when R_0 is distinctly above one was known in 2001, but it was restricted to the body of the distribution where $y_1(n) = O(1)$. We find therefore that the extension to the near-end tails described by the

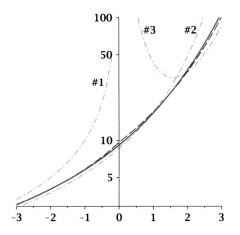

Fig. 15.3 The expected time to extinction from the quasi-stationary distribution $E\tau_Q$ is shown as a function of $\rho = (R_0 - 1)\sqrt{N}$. Comparisons are made with four different explicit approximations, namely the uniform approximation (shown *dashed*), the approximation for $R_0 < 1$ (denoted #1), the approximation in the transition region (denoted #2), and the approximation for $R_0 > 1$ (denoted #3). The uniform approximation appears superior to the other three. The vertical scale is logarithmic, $N = 100$, and $\mu = 1$

condition $y_1(n) = o(\sqrt{N})$ is new. Similarly, the approximation of the left tail of the quasi-stationary distribution given in (11.2) when R_0 is distinctly above one and $n = o(N)$ is new, while the approximation in the left-most tail in (11.4) for $n = o(\sqrt{N})$ was known in 2001.

Another improvement over the results that were available in 2001 lies in the fact that we have derived approximations that are uniform over the three parameter regions. There are two advantages with this improvement. Both of them are easy to understand for the expected time to extinction from quasi-stationarity, $E\tau_Q$, illustrated in Fig. 15.3, but they hold equally well for the quasi-stationary distribution. The first advantage has already been commented on with reference to Fig. 15.3, namely that the uniform approximation gives a better approximation than either of the approximations whose validity is restricted to one of the three parameter regions. The second advantage is that the availability of a uniform approximation leads directly to a unique approximation for any given value of R_0 or ρ, while there are ranges of parameter values where the choice between available nonuniform approximations is unclear.

The stochastic ordering that was studied in Chap. 3 has led to the rather satisfactory result that we have found asymptotic approximations of the quasi-stationary distribution in the two parameter regions where R_0 is distinctly different from one. This is clearly a much more satisfactory situation than the earlier one where we derived asymptotic approximations of iterates under the map Ψ. The fact that we have not been able to derive an asymptotic approximation of the quasi-stationary distribution in the transition region does not diminish the advantage with this approach.

The logistic SIS model is probably the simplest model that one can imagine for endemic infections, and that contains the important features that it accounts for density dependence and that its deterministic version is equipped with a threshold. As indicated in Chap. 2, there are many other logistic models for which the stochastic analysis is awaiting to be carried out. We have described the analysis of the SIS model in a way that should guide the analysis of the quasi-stationary distribution of other models.

We mention finally that several open problems have been identified along the road. Solution of these problems may help in analyzing other models, or in deriving improved approximations of the SIS model.

Appendix A
Notation

We summarize here the notation that is used for describing the approximations of the quasi-stationary distribution q and of the stationary distributions $p^{(1)}$ and $p^{(0)}$ of the two auxiliary processes of the stochastic SIS model. All quantities included are expressed in terms of the two essential parameters N and R_0.

The first quantities to be defined have the subscript 1. They are important for describing results when R_0 is distinctly above one. The quantities $\bar{\mu}_1$ and $\bar{\sigma}_1$ then serve as mean and standard deviation, respectively, of the approximation of the body and the near-end tails of the quasi-stationary distribution. Furthermore, the expected time to extinction from quasi-stationarity then grows exponentially with N. Its logarithm is asymptotically proportional to $\gamma_1 N$. The quantities γ_1 and β_1 are related since $\beta_1^2 = 2\gamma_1 N$. This implies that $\varphi(\beta_1) = \exp(-\gamma_1 N)/\sqrt{2\pi}$. Note also that $\gamma_1 \geq 0$ for $R_0 > 0$. We have

$$\bar{\mu}_1 = \frac{R_0 - 1}{R_0} N, \tag{A.1}$$

$$\bar{\sigma}_1 = \sqrt{\frac{N}{R_0}}, \tag{A.2}$$

$$y_1(n) = \frac{n - \bar{\mu}_1}{\bar{\sigma}_1}, \tag{A.3}$$

$$z_1(n) = \frac{n - \bar{\mu}_1}{\bar{\sigma}_1^2}, \tag{A.4}$$

$$\beta_1 = \text{sign}(R_0 - 1)\sqrt{2N\left(\log R_0 - \frac{R_0 - 1}{R_0}\right)}, \tag{A.5}$$

$$\gamma_1 = \log R_0 - \frac{R_0 - 1}{R_0}. \tag{A.6}$$

I. Nåsell, *Extinction and Quasi-stationarity in the Stochastic Logistic SIS Model*, Lecture Notes in Mathematics 2022, DOI 10.1007/978-3-642-20530-9, © Springer-Verlag Berlin Heidelberg 2011

The next quantities have the subscript 2. They appear in approximations of the left tail of the quasi-stationary distribution when R_0 is distinctly above the value one. They are

$$\bar{\mu}_2 = N \log R_0, \tag{A.7}$$

$$\bar{\sigma}_2 = \sqrt{N}, \tag{A.8}$$

$$y_2(n) = \frac{n - \bar{\mu}_2}{\bar{\sigma}_2}, \tag{A.9}$$

$$z_2(n) = \frac{n - \bar{\mu}_2}{\bar{\sigma}_2^2}, \tag{A.10}$$

$$\beta_2 = \sqrt{N} \log R_0. \tag{A.11}$$

The study of the transition region requires a rescaling of R_0 with the aid of a quantity ρ defined as follows:

$$\rho = (R_0 - 1)\sqrt{N}. \tag{A.12}$$

The transition region is characterized by the requirement that $\rho = O(1)$ as $N \to \infty$.

The next quantities have the subscript 3. They are useful in describing the quasi-stationary distribution in the transition region:

$$\bar{\mu}_3 = \rho\sqrt{N}, \tag{A.13}$$

$$\bar{\sigma}_3 = \bar{\sigma}_2 = \sqrt{N}, \tag{A.14}$$

$$y_3(n) = \frac{n - \bar{\mu}_3}{\bar{\sigma}_3}, \tag{A.15}$$

$$z_3(n) = \frac{n - \bar{\mu}_3}{\bar{\sigma}_3^2}, \tag{A.16}$$

$$\beta_3 = \frac{R_0 - 1}{R_0}\sqrt{N}. \tag{A.17}$$

It is useful to note that $\rho = 0$ when $R_0 = 1$, and that ρ has the same sign as $R_0 - 1$. The quantities $\beta_1, \beta_2, \beta_3$ have similar behaviors. They deviate slightly from each other in the transition region, as shown by the following asymptotic approximations:

$$\beta_1 \sim \rho - \frac{2}{3}\frac{\rho^2}{\sqrt{N}}, \quad \rho = O(1), \quad N \to \infty, \tag{A.18}$$

$$\beta_2 \sim \rho - \frac{1}{2}\frac{\rho^2}{\sqrt{N}}, \quad \rho = O(1), \quad N \to \infty, \tag{A.19}$$

$$\beta_3 \sim \rho - \frac{\rho^2}{\sqrt{N}}, \quad \rho = O(1), \quad N \to \infty. \tag{A.20}$$

The description of the uniform results in Chap. 12 uses the following notation:

$$\bar{\mu}_{12} = \begin{cases} \bar{\mu}_1, & R_0 \geq 1, \\ \bar{\mu}_2, & R_0 \leq 1, \end{cases} \tag{A.21}$$

$$\bar{\sigma}_{12} = \begin{cases} \bar{\sigma}_1, & R_0 \geq 1, \\ \bar{\sigma}_2, & R_0 \leq 1, \end{cases} \tag{A.22}$$

$$y_{12}(n) = \begin{cases} y_1(n), & R_0 \geq 1, \\ y_2(n), & R_0 \leq 1, \end{cases} \tag{A.23}$$

$$r_0 = \begin{cases} \sqrt{R_0}, & R_0 \geq 1, \\ R_0, & R_0 \leq 1, \end{cases} \tag{A.24}$$

$$\beta_0 = \rho, \tag{A.25}$$

and

$$\beta_{ij} = \begin{cases} \beta_i, & R_0 \geq 1, \quad i = 0, 1, 2, 3, \\ \beta_j, & R_0 \leq 1, \quad j = 0, 1, 2, 3. \end{cases} \tag{A.26}$$

We note that all of the quantities $\beta_0 = \rho$, β_1, β_2, and β_3 are asymptotic to $\rho = (R_0 - 1)\sqrt{N}$ in the transition region. Also, all of them are equal to zero when $\rho = 0$. Thus, all of the quantities β_{ij} are continuous functions of ρ. They are equal to zero when $\rho = 0$, and they all have the same sign as ρ. This means that they are positive for $R_0 > 1$ and negative for $R_0 < 1$.

The functions G_1, G_2, and G are defined as follows:

$$G_1(z) = \begin{cases} 1 - \dfrac{1}{z} + \dfrac{1}{\exp(z) - 1}, & z \neq 0, \\ \dfrac{1}{2}, & z = 0, \end{cases} \tag{A.27}$$

$$G_2(x, z) = \begin{cases} \dfrac{1 - \exp(-xz)}{z}, & z \neq 0, \\ x, & z = 0. \end{cases} \tag{A.28}$$

$$G(z) = \begin{cases} \dfrac{1}{z} \log \dfrac{\exp(z) - 1}{z}, & z \neq 0, \\ \dfrac{1}{2}, & z = 0. \end{cases} \tag{A.29}$$

The functions H_1, H_1^*, H_0, and H are defined as follows:

$$H_1(y) = \frac{\Phi(y)}{\varphi(y)}, \tag{A.30}$$

$$H_1^*(y) = \frac{\Phi^*(y)}{\varphi^*(y)}, \tag{A.31}$$

$$H_0(y) = \begin{cases} H_a(y), & y \le \rho_b, \\ H_a(\rho_b) + \displaystyle\int_{\rho_b}^{y} H_1(t)\,dt, & \rho > \rho_b, \end{cases} \tag{A.32}$$

where the auxiliary function H_a is defined by

$$H_a(y) = -\log|y| + \sum_{k=1}^{m_d} (-1)^k \frac{a_k}{2k} \frac{1}{y^{2k}}, \quad y < 0, \tag{A.33}$$

and m_d is a positive integer depending on ρ_b. The Maple procedure in Appendix B uses $\rho_b = -3$ and $m_d = 4$. The coefficients a_k are defined by

$$a_k = \frac{(2k)!}{k!2^k}, \quad k = 0, 1, 2, \dots. \tag{A.34}$$

Finally, the functions H, $\tilde\rho$, and R are defined as follows:

$$H(y) = \frac{1}{y + 1/H(y)} \int_{-1/H(y)}^{y} H_1(t)\,dt, \tag{A.35}$$

$$\tilde\rho(y) = y + \frac{1}{H(y)}, \tag{A.36}$$

and

$$R(N, y) = 1 + \frac{\tilde\rho(y)}{\sqrt{N}} = 1 + \frac{y + 1/H(y)}{\sqrt{N}}. \tag{A.37}$$

Appendix B
A Maple Module for Numerical Evaluations

The numerical evaluations that are reported in this monograph have all been done with Maple. We have used the Maple procedures that are included in the Maple module named SIS given below. The module contains 21 procedures. We give some brief comments on them.

We begin with six scalar functions. The normal distribution function Φ and the normal density function φ are evaluated with the Maple functions **PHI** and **phi**, respectively. Similarly, the three functions H, H_0, and H_1 are evaluated using the Maple functions **H**, **H0**, and **H1**, respectively. In addition, the Maple procedure **h** is used to determine **H**. It calls **H1**.

The remaining 15 procedures all deal with probability vectors. These probability vectors are treated in the form of one-dimensional Maple Arrays. With the exception of the last two procedures (**ratio** and **short**), the lengths of the vectors are N, the population size for the SIS model. The first four of these procedures are **PSI**, **PSImn**, **rf**, and **cdf**. The procedure **PSI** is used to determine the image under the map Ψ of some vector p. A related procedure is the one called **PSImn**. It is used to determine the image of the vector $p^{(1)}$ or $p^{(0)}$ after application of the map Ψ n times. The ratio r_k defined in (3.35) can be studied with the procedure **rf**. It is called by the procedure **PSI**. Another useful procedure is called **cdf**. It converts a probability vector p into its distribution function. It is called by both **rf** and by **PSI**.

There are three procedures associated with each of the two stationary distributions $p^{(0)}$ and $p^{(1)}$, and three more associated with the quasi-stationary distribution q. First of all, each of these distributions can be evaluated numerically with the aid of the procedures called **p0f**, **p1f**, and **qsf**, respectively. Furthermore, we note that uniform approximations of body and of left tail of these distributions are given in Chap. 13. Procedures that evaluate these approximations numerically are contained in **p0bodyappr**, **p0tailappr**, **p1bodyappr**, **p1tailappr**, **qsbodyappr**, and **qstailappr**, respectively.

In addition, the module contains two procedures that can be useful in certain circumstances. They are called **ratio** and **short**. The first of these takes a vector of length M with components $p_1, p_2, \ldots, p_M$, and produces a vector of length $M-1$

whose component k equals the ratio p_{k+1}/p_k. The last procedure shortens a Maple Array to its first M elements. It can be useful in plotting distributions.

We conclude this appendix by a listing of the Maple module named SIS:

```
SIS:=module()
  export PHI,phi,H,H0,H1,h,PSI,PSImn,rf,cdf,
    p0f,p1f,qsf,p0bodyappr,p0tailappr,p1bodyappr,
    p1tailappr,qsbodyappr,qstailappr,ratio,short:
  option package;
  description "The qsd of the SIS model, and related
    quantities";

  PHI:=y->(1+erf(y/sqrt(2)))/2;

  phi:=y->exp(-y^2/2)/sqrt(2*Pi):

  H:=x->1/fsolve(h(x,v)=1,v):

  H0:=proc(x)
    local y;
    if evalf(x)<= -3 then
      evalf(-log(abs(x))-1/2/x^2+3/4/x^4-15/6/x^6
          +105/8/x^8);
    else
      evalf(-log(3)-835/17496 + Int(H1(y),y=-3..x));
    fi;
  end proc:

  H1:=x->PHI(x)/phi(x):

  h:=proc(x,v)
    local y;
    if v=-x then
      -x*H1(x);
    else
      (v/(x+v))*evalf(Int(H1(y),y=-v..x));
    fi;
  end proc:

  PSI:=proc(N,R0,p)
    local p0,pin,r,S,pp,ppsum:
    description ''Determine the image under the map PSI
      of the vector p'':
    p0:=p0f(N,R0):
```

```
    pin:=p0/p0[1]:
    r:=rf(N,R0,p):
    S:=cdf(r):
    pp:=pin*S:
    ppsum:=add(pp[k],k=1..N):
    pp:=pp/ppsum:
  end proc:


PSImn:=proc(N,R0,m,n)
  local p0,p1,pin,p,k:
  description ''Determines PSI^n(p^(m)), m=0 or 1'':
  p0:=p0f(N,R0):
  p1:=p1f(N,R0):
  pin:=p0/p0[1]:
  if m=0 then
    p:=p0:
  else
    p:=p1:
  fi:
  if n>0 then
    for k from 1 to n do
      p:=PSI(N,R0,p):
    od:
  fi:
  p:
end proc:


rf:=proc(N,R0,p)
  local p1,rho,pcum,rr,k:
  description ''The ratio r with vector p in
    the numerator'':
  p1:=p1f(N,R0):
  rho:=p1/p1[1]:
  pcum:=cdf(p):
  rr:=Array(1..N,1):
  for k from 2 to N do
    rr[k]:=(1-pcum[k-1])/rho[k]:
  od:
  rr:
end proc:


cdf:=proc(p)
  local F,N,k;
  description "The sum of p[j] from 1 to k":
```

```
  N:=op(2,ArrayDims(p));
  F:=Array(1..N);
  F[1]:=p[1];
  for k from 2 to N do
    F[k]:=F[k-1]+p[k];
  od;
  F;
end proc;


p0f:=proc(N,R0)
  local p0,k0,k,sump0;
  description "The stationary distribution p0":
  p0:=Array(1..N,1.);
  k0:=max(1,floor((R0-1)*N/R0));
  for k from k0+1 to N do
    p0[k]:=p0[k-1]*(1-1/k)*(1-(k-1)/N)*R0;
  od;
  for k from k0-1 by -1 to 1 do
    p0[k]:=p0[k+1]*(1+1/k)/R0/(1-k/N);
  od;
  sump0:=add(p0[k],k=1..N):
  p0:=p0/sump0;
  p0:
end proc:


p1f:=proc(N,R0)
  local p1,k0,k,sump1;
  description "The stationary distribution p1":
  p1:=Array(1..N,1.):
  k0:=max(1,floor((R0-1)*N/R0));
  for k from k0+1 to N do
    p1[k]:=p1[k-1]*(1-(k-1)/N)*R0;
  od;
  for k from k0-1 by -1 to 1 do
    p1[k]:=p1[k+1]/R0/(1-k/N);
  od;
  sump1:=add(p1[k],k=1..N):
  p1:=p1/sump1:
end proc:


qsf:=proc(N,R0)
  local qs1,kvot,qs2,kvotmax,kvotmin,kv:
  description "The quasi-stationary distribution":
  if R0>1 then
```

```
     qs1:=p0f(N,R0):
   else
     qs1:=p1f(N,R0):
   fi:
   kv:=2:
   while kv > 1+1e-6 do
   qs2:=PSI(N,R0,qs1):
   kvot:=qs1/qs2:
   kvotmax:=max(kvot):
   kvotmin:=min(kvot):
   kv:=max(kvotmax,1/kvotmin):
   qs1:=qs2:
   od:
   qs2:
   end proc:


p0bodyappr:=proc(N,R0)
   local rho,beta2,beta3,mu1,sigma1,mu2,sigma2,a,
     beta30,beta32,mu12,sigma12,r0,aa,f,p0:
   description ``Approximation of the body of p0'':
   rho:=(R0-1)*sqrt(N):
   beta2:=sqrt(N)*log(R0):
   beta3:=(R0-1)*sqrt(N)/R0:
   mu1:=(R0-1)*N/R0: sigma1:=sqrt(N/R0):
   mu2:=N*log(R0): sigma2:=sqrt(N):
   if R0>1 then
     r0:=1: beta30:=beta3: beta32:=beta3: mu12:=mu1:
       sigma12:=sigma1:
   else
     r0:=R0: beta30:=rho: beta32:=beta2: mu12:=mu2:
       sigma12:=sigma2:
   fi:
   a:=sqrt(R0/r0):
   aa:=evalf(a/(0.5*log(N)+H0(beta30))/phi(beta32)):
   f:=n->aa*evalf(phi((n-mu12)/sigma12)/n):
   p0:=Array(1..N,f):
   end proc:


p0tailappr:=proc(N,R0)
   local rho,beta1,beta2,beta3,mu2,sigma2,a,beta10,
     aa,f,p0:
   description ``Approximation of the left tail of
     p0'':
   rho:=(R0-1)*sqrt(N):
```

```
    beta1:=sqrt(2*N*(log(R0)-(R0-1)/R0)):
    beta2:=sqrt(N)*log(R0):
    beta3:=(R0-1)*sqrt(N)/R0:
    mu2:=N*log(R0): sigma2:=sqrt(N):
    if R0>1 then
      a:=beta3/beta1: beta10:=beta1:
    else
      a:=1: beta10:=rho:
    fi:
    aa:=evalf(a/(0.5*log(N)+H0(beta10))/phi(beta2)):
    f:=n->aa*evalf(phi((n-mu2)/sigma2)/n):
    p0:=Array(1..N,f):
  end proc:


p1bodyappr:=proc(N,R0)
    local beta1,beta2,beta3,mu1,sigma1,mu2,sigma2,
      beta12,beta13,mu12,sigma12,a,f,p1:
    description ``Approximation of the body of p1'':
    beta1:=sqrt(2*N*(log(R0)-(R0-1)/R0)):
    beta2:=sqrt(N)*log(R0):
    beta3:=(R0-1)*sqrt(N)/R0:
    mu1:=(R0-1)*N/R0: sigma1:=sqrt(N/R0):
    mu2:=N*log(R0): sigma2:=sqrt(N):
    if R0>1 then
      beta12:=beta1: beta13:=beta1: mu12:=mu1:
        sigma12:=sigma1:
    else
      beta12:=beta2: beta13:=beta3: mu12:=mu2:
        sigma12:=sigma2:
    fi:
    a:=evalf(1/H1(beta13)/sigma12/phi(beta12)):
    f:=n->a*evalf(phi((n-mu12)/sigma12)):
    p1:=Array(1..N,f):
    end proc:


p1tailappr:=proc(N,R0)
    local beta1,beta2,beta3,mu2,sigma2,beta13,a,f,p1:
    description ``Approximation of the left tail of
      p1'':
    beta1:=sqrt(2*N*(log(R0)-(R0-1)/R0)):
    beta2:=sqrt(N)*log(R0):
    beta3:=(R0-1)*sqrt(N)/R0:
    mu2:=N*log(R0): sigma2:=sqrt(N):
    if R0>1 then
```

```
      beta13:=beta1:
   else
      beta13:=beta3:
   fi:
   a:=evalf(1/H1(beta13)/sigma2/phi(beta2)):
   f:=n->a*evalf(phi((n-mu2)/sigma2)):
   p1:=Array(1..N,f):
end proc:


qsbodyappr:=proc(N,R0)
   local rho,R,beta2,mu1,sigma1,mu2,sigma2,beta02,r0,
      mu12,sigma12,a,aa,f,qs:
   description ''Approximation of the body of qs'':
   rho:=(R0-1)*sqrt(N):
   R:=evalf(1+(rho+1/H(rho))/sqrt(N)):
   beta2:=sqrt(N)*log(R0):
   mu1:= (R0-1)*N/R0: sigma1:=sqrt(N/R0):
   mu2:=N*log(R0): sigma2:=sqrt(N):
   if R0>1 then
      r0:=1: beta02:=rho: mu12:=mu1: sigma12:=sigma1:
   else
      r0:=R0: beta02:=beta2: mu12:=mu2: sigma12:=sigma2:
   fi:
   a:=1/sqrt(R0*r0):
   aa:=evalf(a/(1+rho*H(rho))/phi(beta02)):
   f:=n->aa*evalf(phi((n-mu12)/sigma12)*(1-1/R^n)/n):
   qs:=Array(1..N,f):
   end proc:


qstailappr:=proc(N,R0)
   local rho,R,beta1,beta2,beta10,mu2,sigma2,
      a,aa,f,qs:
   description ''Approximation of the left tail of
      qs'':
   rho:=(R0-1)*sqrt(N):
   R:=evalf(1+(rho+1/H(rho))/sqrt(N)):
   beta1:=sqrt(2*N*(log(R0)-(R0-1)/R0)):
   beta2:=sqrt(N)*log(R0):
   mu2:=N*log(R0): sigma2:=sqrt(N):
   if R0>1 then
      a:=rho/R0/beta1: beta10:=beta1:
   else
      a:=1/R0: beta10:=rho:
   fi:
```

```
  aa:=evalf(a/(1+beta10*H(*beta10))/phi(beta2)):
  f:=n->aa*evalf(phi((n-mu2)/sigma2)*(1-1/R^n)/n):
  qs:=Array(1..N,f):
end proc:

ratio:=proc(p)
  local rat,k,M;
  description "The ratio p[k+1]/p[k]":
  M:=op(2,ArrayDims(p));
  rat:=Array(1..M-1):
  for k from 1 to M-1 do
    rat[k]:=p[k+1]/p[k];
  od;
  rat:
end proc;

short:=proc(p,M)
  local sh,k:
  description "Shortens the Array p to its first M
    elements":
  sh:=Array(1..M):
  for k from 1 to M do
    sh[k]:=p[k]:
  od:
  sh:
end proc:

end module:
```

References

1. M. Abramowitz and I. A. Stegun, *Handbook of Mathematical Functions*, Dover, New York (1965).
2. L. J. S. Allen, *An Introduction to Stochastic Processes with Applications to Biology*, Pearson Education Inc., Upper Saddle River, N.J., 2003.
3. H. Andersson and T. Britton, *Stochastic Epidemic Models and Their Statistical Analysis*, Lecture Notes in Statistics **151**, Springer Verlag, New York, 2000.
4. H. Andersson and B. Djehiche, *A threshold limit theorem for the stochastic logistic epidemic*, J. Appl. Prob. **35** (1998), 662–670.
5. T.M. Apostol, *An elementary view of Euler's summation formula*, American Mathematical Monthly, **106**, no 5 (1999), 409-418.
6. M. Assaf and B. Meerson, *Spectral theory of metastability and extinction in birth-death systems*, Physical Review Letters, **97** (2006).
7. N. T. J. Bailey, *The Mathematical Theory of Infectious Diseases*, Charles Griffin & Company Ltd, London and High Wycombe, 1975.
8. A. D. Barbour, *Quasi-stationary distributions in Markov population processes*, Adv. Appl. Prob. **8** (1976), 296–314.
9. D. J. Bartholomew, *Continuous time diffusion models with random duration of interest*, J. Math. Sociol. **4** (1976), 187–199.
10. M. S. Bartlett, *Deterministic and stochastic models for recurrent epidemics*, In: Proceedings of the Third Berkeley Symposium on Mathematics, Statistics and Probability, vol 4, University of Calofornia Press, Berkeley, (1956), 81–109.
11. M. S. Bartlett, *On theoretical models for competitive and predatory biological systems*, Biometrika **44** (1957), 27–42.
12. M. S. Bartlett, *The critical community soze fopr measles in the United States*, J. Roy. Stat. Soc. Ser. A **123** (1960), 37–44.
13. M. S. Bartlett, J. C. Gower, and P. H. Leslie, *A comparison of theoretical and empirical results for some stochastic population models*, Biometrika **47** (1960), 1–11.
14. A. T. Bharucha-Reid, *Elements of the Theory of Markov Processes*, New York: McGraw-Hill, 1960.
15. B. J. Cairns, J. V. Ross, and T. Taimre, *A comparison of models for predicting population persistence*, Ecological Modelling **201** (2007), 19–26.
16. J. A. Cavender, *Quasi-stationary distributions of birth-and-death processes*, Adv. Appl. Prob. **10** (1978), 570—586.
17. M. F. Chen, *Eigenvalues, Inequalities, and Ergodic Theorry*, Springer, London, 2005.
18. D. Clancy and P. K. Pollett, *A note on quasi-stationary distributions of birth-death processes and the SIS logistic epidemic*, J. Appl. Prob. **40** (2003), 821–825.

19. D. Clancy and S. T. Mendy, *Approximating the quasi-stationary distribution of the SIS model for endemic infection*, Methodol. Comput. Appl. Probab. (2010).

20. D. R. Cox, *The continuity correction*, Biometrika, **57** No 1 (1970), 217–219.

21. J.N. Darroch and E. Sneta, *On quasi-stationary distributions in absorbing continuous-time finite Markov chains*, J. Appl Prob. **4** (1967), 192–196.

22. N.G. de Bruijn, *Asymptotic Methods in Analysis*, Wiley, New York, 1957.

23. R. Dickman and R. Vidigal, *Quasi-stationary distributions for stochastic processes with an absorbing state*, J. Phys. A: Math. Gen. **35** (2002), 1147–1166.

24. C.R. Doering, K.V. Sargsyan and L.M. Sander, *Extinction times for birth-death processes: Exact results, continuum asymptotics, and the failure of the Fokker-Planck approximation*, SIAM J. Multiscale Modeling and Sim. **3** (2005), 283–299.

25. R.G. Dolgoarshinnykh and S.P. Lalley, *Critical scaling for the SIS stochastic epidemic*, J. Appl. Prob., **43** (2006), 892–898.

26. W. Feller, *Die Grundlagen des Volterraschen Theorie des Kampfes ums Dasein in wahrschein-lichkeitstheoretischer Behandlung*, Acta Biotheoretica, **5** (1939), 11–40.

27. P. Ferrari, H. Kesten, S. Martínez, and P. Picco, *Existence of quasi-stationary distributions. A renewal dynamic approach*, Ann. Probab. **23** (1995), 501–521.

28. P. Flajolet, P. J. Grabner, P. Kirschenhofer, and H. Prodinger *On Ramanujan's Q-function* J. Comp. Appl. Math. **58** (1995), 103–116.

29. J.-P. Gabriel, F. Saucy, and L.-F. Bersier, *Paradoxes in the logistic equation?*, Ecological Modelling, **85** (2005), 147–151.

30. C. W. Gardiner, *Handbook of Stochastic Methods for Physics, Chemistry, and the Natural Sciences*, Springer Verlag, Berlin, Heidelberg, New York, London, Paris, Tokyo, Hong Kong, Barcelona, Budapest, 1985.

31. J. Grasman, *Stochastic epidemics: The expected duration of the endemic period in higher dimensional models*, Math. Biosci. **152** (1998), 13–27.

32. T. E. Harris, *Contact interactions on a lattice*, Ann. Prob. **2**, 969–988, 1974.

33. M. Iosifescu and P. Tautu, *Stochastic Processes and Applications in Biology and Medicine*, Vol. II. Berlin: Springer Verlag, 1973.

34. S. Karlin and H. M. Taylor, *A First Course in Stochastic Processes*, Academic Press, New York, 1975.

35. D. G. Kendall, *Stochastic processes and population growth*, J. Roy. Stat. Soc. Series B **11**, 230–264, 1949.

36. M. Kijima, *Markov Processes for Stochastic Modeling*, Chapman & Hall, London, 1997.

37. M. Kijima and E. Seneta, *Some Results for Quasi-Stationary Distributions of Birth-Death Processes*, J. Appl. Prob. **28** No. 3 (1991), 503–511.

38. D. E. Knuth, *The Art of Computer Programming. Volume 1: Fundamental Algorithms*, Addison-Wesley, Reaading, MA, 1973.

39. R. J. Kryscio and C. Lefèvre, *On the extinction of the SIS stochastic logistic epidemic*, J. Appl. Prob. **27** (1989), 685–694.

40. J.H. Matis and T.R. Kiffe, *On approximating the moments of the equilibrium distribution of a stochastic logistic model*, Biometrics **52** No 3 (1996), 980–991.

41. R. M. May, *Stability and Complexity in Model Ecosystems*, Princeton University Press, Princeton, N.J. 1974.

42. I. Nåsell, *On the quasi-stationary distribution of the Ross malaria model*, Math. Biosci. **107** No 2 (1991), 187–207.

43. I. Nåsell, *The threshold concept in stochastic epidemic and endemic models*, In Epidemic Models: Their Structure and Relation to Data (D. Mollison, ed.), Publications of the Newton Institute, Cambridge University Press, Cambridge, 1995, 71–83.

44. I. Nåsell, *The quasi-stationary distribution of the closed endemic SIS model*, Adv. Appl. Prob. **28** (1996), 895–932.

45. I. Nåsell, *On the quasi-stationary distribution of the stochastic logistic epidemic*, Math. Biosci. **156** (1999a), 21–40.

46. I. Nåsell, *On the time to extinction in recurrent epidemics*, J. Roy. Statist. Soc. Series B **61** (1999b), 309–330.

47. I. Nåsell, *Extinction and quasi-stationarity in the Verhulst logistic model*, J. Theor. Biol. **211** (2001), 11–27.

48. I. Nåsell, *Endemicity, persistence, and quasi-stationarity*, Mathematical Approaches for Emerging and Reemerging Infectious Diseases, (C Castillo-Chavez, S. Blower, P. van den Driessche, D. Kirschner, A. Yakubu, eds.), The IMA Volumes in Mathematics and its Applications, Volume 125, Springer Verlag, New York, 2002, 199–227.

49. I. Nåsell, *Moment closure and the stochastic logistic model*, Theor. Pop. Biol. **63** (2003) 159–168.

50. I. Nåsell, *A new look at the critical community size for childhood infections*, Theor. Pop. Biol. **67** (2005) 203–216.

51. T.J. Newman and J-B Ferdy and C Quince, *Extinction times and moment closure in the stochastic logistic process*, Theor. Pop. Biol. **65** (2004), 115–126.

52. R. M. Nisbet and W. S. C. Gurney, *Modelling Fluctuating Populations*, Wiley, New York, 1982.

53. R. H. Norden, *On the distribution of the time to extinction in the stochastic logistic ppopulation model*, Adv. Appl. Prob. **14** (1982), 687–708.

54. F. W. J. Olver, *Asymptotics and Special Functions*, Academic Press, New York, San Fransisco, London, 1974.

55. F. W. J. Olver, D. W. Lozier, R. F. Boisvert, and C. W. Clark, eds., *NIST Handbook of Mathematical Functions*, http://dlmf.nist.gov, 2010.

56. I. Oppenheim, K. E. Shuler, and G. H. Weiss, *Stochastic theory of nonlinear rate processes with multiple stationary states*, Physica **88A** (1977), 191–214.

57. O. Ovaskainen, *The quasistationary distribution of the stochastic logistic model*, J. Appl. Prob., **38** (2001), 898–907.

58. L. Pearl and L. J. Reed, *On the rate of growth of the population of the United States since 1790 and its mathematical representation*, Proc. Natl. Acad. of Sci. USA **6**, 275–288, 1920.

59. P. K. Pollett, *Reversibility, invariance, and μ-invariance*, Adv. Appl. Prob. **20** (1988), 600–621.

60. P. K. Pollett, *The generalized Kolmogorov criterion*, Stoch. Proc Appl. **33** (1989), 29–44.

61. P. K. Pollett, *Quasi-Stationary Distributions: A Bibliography*, http://www.maths.uq.edu.au/~pkp/papers/qsds/qsds.pdf, 2010.

62. B. J. Prendiville, *Discussion on symposium on stochastic processes*, J. Roy. Stat. Soc. Ser. B **11**, 273, 1949.

63. S. Ramanujan, *Question 294*, J. Indian Math. Soc. **3** (1911), 128.

64. E. Renshaw, *Modelling Biological Populations in Space and Time*, Cambridge University Press, Cambridge, 1991.

65. L. M. Ricciardi, *Stochastic population theory: birth and death processes*, Mathematical Ecology, an Introduction (Hallam & Levin, eds.) **17**, 155–190, 1980, Berlin: Springer Verlag.

66. M. Schmitz, *Quasi-stationarität in einem epidemiologischen Modell*, http://wwwmath.uni-muenster.de/statistik/alsmeyer /Diplomarbeiten/Schmitz_Manuela.pdf, 2006.

67. E. Seneta, *Quasi-stationary behavior in the random walk with continuous time*, Austr. J. Statist. **8** (1966), 92–98.

68. A. Singh and J.P. Hespanha, *A derivative matching approach to moment closure for the stochastic logistic model*, Bull. Math. Biol. **69** no6 (2007), 1909–1925.

69. D. Sirl, H. Zhang, and P. Pollett, *Computable bounds for the decay parameter of a birth-death process*, J. Appl. Prob. **44** (2007), 476–491.

70. M. Takashima, *Note on evolutionary processes*, Bull. Math. Stat. **7**, 18–24, 1957.

71. H. M. Taylor and S. Karlin, *An Introduction to Stochastic Modeling*, Academic Press, San Diego, 1993.

72. A. Tsoularis and J. Wallace, *Analysis of logistic growth models*, Math. Biosci. **179** (2002), 21–55.

73. E. A. van Doorn, *Quasi-stationary distributions and convergence to quasi-stationarity of birth-death processes*, Adv. Appl. Prob. **23** (1991), 683–700.

74. P. F. Verhulst, *Notice sur la loi que la population suit dans son accroisement*, Corr. Math. Phys. **X**, 113–121, 1838.

75. P. F. Verhulst, *Recherches mathématiques sur la loi d'accroissement de la population*, Noveau Mémoires de l'Academie Royale des Sciences et Belles-Lettres de Bruxelles, **18**, 1–41, 1845.

76. G. H. Weiss and M. Dishon, *On the asymptotic behavior of the stochastic and deterministic models of an epidemic*, Math. Biosci. **11** (1971), 261–265.

77. P. Whittle, *On the use of the normal approximation in the treatment of stochastic processes*, J. Roy. Stat. Soc. Ser B **19**, 268–281, 1957.

78. A. M. Yaglom, *Certain limit theorems of the theory of branching processes (in Russian)*, Dokl. Akad. Nauk SSSR **56** (1947), 795–798.

Index

LECTURE NOTES IN MATHEMATICS

Edited by J.-M. Morel, B. Teissier, P.K. Maini

Editorial Policy (for the publication of monographs)

1. Lecture Notes aim to report new developments in all areas of mathematics and their applications - quickly, informally and at a high level. Mathematical texts analysing new developments in modelling and numerical simulation are welcome.

   Monograph manuscripts should be reasonably self-contained and rounded off. Thus they may, and often will, present not only results of the author but also related work by other people. They may be based on specialised lecture courses. Furthermore, the manuscripts should provide sufficient motivation, examples and applications. This clearly distinguishes Lecture Notes from journal articles or technical reports which normally are very concise. Articles intended for a journal but too long to be accepted by most journals, usually do not have this "lecture notes" character. For similar reasons it is unusual for doctoral theses to be accepted for the Lecture Notes series, though habilitation theses may be appropriate.

2. Manuscripts should be submitted either online at www.editorialmanager.com/lnm to Springer's mathematics editorial in Heidelberg, or to one of the series editors. In general, manuscripts will be sent out to 2 external referees for evaluation. If a decision cannot yet be reached on the basis of the first 2 reports, further referees may be contacted: The author will be informed of this. A final decision to publish can be made only on the basis of the complete manuscript, however a refereeing process leading to a preliminary decision can be based on a pre-final or incomplete manuscript. The strict minimum amount of material that will be considered should include a detailed outline describing the planned contents of each chapter, a bibliography and several sample chapters.

   Authors should be aware that incomplete or insufficiently close to final manuscripts almost always result in longer refereeing times and nevertheless unclear referees' recommendations, making further refereeing of a final draft necessary.

   Authors should also be aware that parallel submission of their manuscript to another publisher while under consideration for LNM will in general lead to immediate rejection.

3. Manuscripts should in general be submitted in English. Final manuscripts should contain at least 100 pages of mathematical text and should always include

   – a table of contents;
   – an informative introduction, with adequate motivation and perhaps some historical remarks: it should be accessible to a reader not intimately familiar with the topic treated;
   – a subject index: as a rule this is genuinely helpful for the reader.

   For evaluation purposes, manuscripts may be submitted in print or electronic form (print form is still preferred by most referees), in the latter case preferably as pdf- or zipped ps-files. Lecture Notes volumes are, as a rule, printed digitally from the authors' files. To ensure best results, authors are asked to use the LaTeX2e style files available from Springer's web-server at:

   ftp://ftp.springer.de/pub/tex/latex/svmonot1/ (for monographs) and
   ftp://ftp.springer.de/pub/tex/latex/svmultt1/ (for summer schools/tutorials).
   Additional technical instructions, if necessary, are available on request from:
   lnm@springer.com.

4. Careful preparation of the manuscripts will help keep production time short besides ensuring satisfactory appearance of the finished book in print and online. After acceptance of the manuscript authors will be asked to prepare the final LaTeX source files and also the corresponding dvi-, pdf- or zipped ps-file. The LaTeX source files are essential for producing the full-text online version of the book (see http://www.springerlink.com/openurl.asp?genre=journal&issn=0075-8434 for the existing online volumes of LNM).

   The actual production of a Lecture Notes volume takes approximately 12 weeks.

5. Authors receive a total of 50 free copies of their volume, but no royalties. They are entitled to a discount of 33.3% on the price of Springer books purchased for their personal use, if ordering directly from Springer.

6. Commitment to publish is made by letter of intent rather than by signing a formal contract. Springer-Verlag secures the copyright for each volume. Authors are free to reuse material contained in their LNM volumes in later publications: a brief written (or e-mail) request for formal permission is sufficient.

Addresses:

Professor J.-M. Morel, CMLA,
École Normale Supérieure de Cachan,
61 Avenue du Président Wilson, 94235 Cachan Cedex, France
E-mail: Jean-Michel.Morel@cmla.ens-cachan.fr

Professor B. Teissier, Institut Mathématique de Jussieu,
UMR 7586 du CNRS, Équipe "Géométrie et Dynamique",
175 rue du Chevaleret,
75013 Paris, France
E-mail: teissier@math.jussieu.fr

For the "Mathematical Biosciences Subseries" of LNM:

Professor P.K. Maini, Center for Mathematical Biology,
Mathematical Institute, 24-29 St Giles,
Oxford OX1 3LP, UK
E-mail: maini@maths.ox.ac.uk

Springer, Mathematics Editorial, Tiergartenstr. 17,
69121 Heidelberg, Germany,
Tel.: +49 (6221) 487-259
Fax: +49 (6221) 4876-8259
E-mail: lnm@springer.com